J.P. McEvoy was awarded an Msc in physics from the University of Pennsylvania and a PhD from the University of London. As a research associate at the RCA Research Laboratories in Princeton and for the next 15 years, he worked in solid state physics as a research scientist in the US, Switzerland and the UK. His two previous books, *Introducing Stephen Hawking* and *Introducing Quantum Theory* have been acclaimed worldwide and translated into a dozen languages. He has also written a popular book on the history of solar eclipses for Fourth Estate. Recently, he has been active in science journalism and broadcasting. JoeMcevoy@mac.com.

D0396037

Titles available in the Brief History series

A BRIEF HISTORY OF

THE UNIVERSE

J.P. McEVOY

ROBINSON RUNNING PRESS
PHILADELPHIA · LONDON

ROBINSON

First published in Great Britain in 2010 by Robinson

5 7 9 10 8 6

Copyright © J.P. McEvoy, 2010

The moral right of the author has been asserted.

A CIP catalogue record for this book
is available from the British Library.

ISBN: 978-1-84529-684-1

Robinson
An imprint of
Little, Brown Book Group
Carmelite House
50 Victoria Embankment
London EC4Y 0DZ

An Hachette UK Company

www.hachette.co.uk
www.littlebrown.co.uk

First Published in the United States in 2010 by Running Press Book Publishers

Books published by Running Press are available at special discounts for bulk purchases in the
United States by corporations, institutions and other organizations. For more information, please
contact the Special Markets Department at the Perseus Books Group, 2300 Chestnut Street,
Suite 200, Philadelphia, PA 19103, or call (800) 810-4145, ext. 5000, or email
special.markets@perseusbooks.com.

US Library of Congress Control Number: 2008931539
US ISBN 978-0-7624-3622-4

10 9 8 7 6 5 4 3 2 1
Digit on the right indicates the number of this printing

Running Press Book Publishers
2300 Chestnut Street
Philadelphia, PA 19103-4371

Visit us on the web!
www.runningpress.com

Printed and bound in Great Britain by CPI Group (UK) Ltd, Croydon CR0 4YY

MIX
Paper from
responsible sources
FSC® C104740

The author dedicates this book to the genius and compassion of Johannes Kepler, the world's first astrophysicist, and Albert Einstein, the world's first theoretical cosmologist . . .

And to my four grandchildren: Emily, Muirenn, Joel and Leelah who have promised me that they will take good care of our tiny part of the universe, the earth.

CONTENTS

PART III: EINSTEIN'S UNIVERSE

LIST OF FIGURES AND ILLUSTRATIONS

Bending of starlight by the Sun. © Author's own sketch.

Possible futures of the universe. © Author's own sketch.

Eddington meets Einstein. © Courtesy of Leiden University.

HR diagram. © Courtesy of George Abell.

Hubble's Law, 1929, 1931. © E. Hubble, M. Humason, 1931.

Raisin bread analogy. © Author's own sketch.

Thermal radiation. © Author's own sketch.

Gamov's letter to Penzias. © *Cosmology and Microwave Astronomy*, A. Penzias, 1972.

ACKNOWLEDGEMENTS

To my wife Patricia, companion for half a century, goes my gratitude for her total acceptance of this project. She gave me the greatest gift a husband could have to follow a dream: a two-year sabbatical from marriage to be a recluse while she used her talent and experience in Ethiopia to help modernize the country's educational system.

I would also like to thank three generous women who helped me during the two years of this book's gestation:
Nicole on the sunny Kent coast for her support and chicken soup; Louisa in wet Glasgow for proofreading and encouragement; and Cynthia in leafy Hampstead for a writer's sanctuary in the final stages.

To my editor and friend, Leo Hollis, I offer my sincere thanks. He deserves most of the credit for the existence of this book. In addition to securing this prize commission for me in the first instance, he has advised, cajoled and creatively criticized my many false starts and 'final' drafts. Over the past two years, as I groped my way through forty centuries of stargazing and theorizing, his erudition, discipline, saintly patience and love of history were indispensable. He instilled in me the confidence that I could encapsulate the main elements of this remarkable story in 100,000 words or so. I probably will never tackle anything like this again. That is, unless Leo Hollis asks me to. But where do we go from here?

I would finally like to acknowledge the continuous support of my wife and three sons, who have never lost faith in me. In order to hold my own in this creative family, I had to do something special. Unlike Bill Starbuck, I'm no good at rainmaking. So I wrote this book about the universe instead.

Special thanks to:
Jenny Doubt for being a painstakingly accurate copy-editor; and Mark McEvoy for preparation of the illustrations under intense deadline pressure albeit in the salubrious surroundings of the Tuscan hills.

'Ptolemy made a universe, which lasted fourteen hundred years.
Newton also made a universe, which has lasted three hundred years.
Einstein has made a universe, and I can't tell you how long that will last.'

George Bernard Shaw
(Introducing Einstein at the Savoy Hotel
in London, 28 October 1930)

PREFACE

Who has not watched nightfall, when the Sun slowly descends below the horizon and, after several minutes, is enveloped in darkness. At that moment the scattering of sunlight by the Earth's atmosphere gives way to a transparent sky. Stars, galaxies, planets and moons, quasars and pulsars appear. Humans have been studying these celestial objects since the beginning of life itself, and it is at this hour that astronomy enthusiasts are poised with their telescopes, binoculars and cameras to study the heavens, visible in all its glory.

Stargazers are joined in a noble and 4,000-year-old ancient tradition of exploring the night sky. This pursuit began as astrology, when our ancestors attempted to describe what they saw in order to read omens and prognostications. This fearful probing of the stars then developed into astronomy, and now has become cosmology – the study of the universe as a whole. Initially, the stargazer's aim was to record the changing face of the skies through the days, months, seasons and years. It was only later that a more complex explanation was sought, and models and plans of how the planets and stars moved in relation to each other were developed.

It was not until the Renaissance that the first scientist began to study these models against what they saw in the firmament and began asking how and why the universe worked? Over the course of a century, knowledge was accumulated by men who have become so famous that their surnames are instantly recognizable: Copernicus, Tycho, Kepler and Galileo. These first modern scientists prioritized observation above received wisdom; experience over theory; and applied mathematics and geometry to theoretical models to reproduce the actual motion of bodies, both heavenly and terrestrial. Most notable for this approach is Isaac Newton, who produced a synthesis of the work of his predecessors, rounding off an era known as the 'scientific revolution'. This critical advance in man's intellectual activity was able to establish principles that were missing from earlier work in science and also to point the way to the extraordinary, abstract concepts to follow in the nineteenth and twentieth centuries, culminating in the work of Einstein.

The study of science had changed, and now astronomical explorers ask not just why and how, but what: Of what were these entities composed? Why did they behave as they did? These questions would demand new ways of looking at the universe, and the development of instruments capable of measuring and recording the heavens. It was Galileo who first pointed his telescope into the night sky and, in so doing, was able to revise what we know about the solar system. This technological revolution would be at the heart of the next era of exploration.

In the British Isles, visionaries like William Herschel and later in 1845, Lord Rosse, built massive versions of this 'optical tube' to look at mysterious celestial objects called nebulae, or clouds. In the next century, inventors applied their skills to extract as much information as possible from the faint beams of light collected by telescopes. Astronomers studied the characteristics of individual stars using new optical techniques such as spectroscopy and photography. With these technological advances, astronomy had become an experimental science.

Still larger telescopes were designed when it became apparent that these methods could determine not only the composition of stars, but also their distance from the Earth and even their speed through the cosmos. American scientific entrepreneurs took up the challenge, squeezing large endowments from private philanthropists to build gigantic machines in the mountains of California.

Finally, observers such as Edwin Hubble and theorists like Einstein joined forces and the golden age of astronomy – the twentieth century – was at hand. Cosmology, the study of the whole universe, was no longer the abstract province of philosophers, and attracted all types of men and women to study the nature and structure of the universe.

This book, which attempts to answer the question of how we discovered the universe, is the story about the individuals whose curiosity, patience and determination have come to characterize the human intellectual spirit. It is an extraordinary story that almost spans the complete history of human civilization. It begins in the ancient society of Babylon between the Tigris and the Euphrates, long considered the cradle of civilization, and continues up to the present day, to the high-tech observatories of the modern world. It is, in effect, the story of science itself.

With such a vast canvas, it has been important to find a narrative through the data. Like the night sky, some stars are brighter than others, and in the history of astronomy this also holds true. It is the story of great people – inventors, theorists, philosophers, noblemen and commoners, librarians and technicians – who have all made their mark on the story of the universe. Stargazing has attracted all types of people, and each of their respective personalities shapes this narrative. Despite their many differences, they all share a determination to venture into the unknown, and return with new knowledge.

Priority has been given to the art of observation, the essential characteristic of all good astronomy. Throughout its 4,000 years, stargazing has revealed new frontiers through combining observations of the heavens with a solid theoretical interpretation of the collected data. One without the other is just bad science – seeing and thinking must continue to go hand in hand.

Thus silent, patient observers still stand at the horizon waiting for darkness. But now those observers are much less confused and fearful, standing upon the shoulders of countless previous generations. This story of how we came to be at home in our universe, is also a plea to continue to seek deeper understanding of our cosmos, our galaxy, our planet, ourselves.

PART I
PTOLEMY'S UNIVERSE

I

THE BABYLONIANS: EARLIEST OBSERVERS OF THE SKY

Two famous rivers, the Tigris and the Euphrates, meet in the area historically known as Mesopotamia, the 'land between two rivers'. Flowing south-eastward, the rivers approach each other to form a single valley before then proceeding in parallel channels for the greater part of their course. They unite again shortly before reaching the Persian Gulf. The delta from these rivers forms a plain about 170 miles (274 kilometres) long. Much like the Nile in Egypt, the delta offered many advantages to early inhabitants, attracting settlements for thousands of years. The fertile valley yielded abundant harvests, workable clay and the nutritious fruit of the date palm. Though large stone deposits were lacking, the early settlers used the local clay for building and writing material.

During the 1870s and early 1880s, numerous clay tablets from Babylonian archaeological sites found their way to antique dealers in Baghdad. The tablets had been found in the ruins of the ancient Assyrian city of Nineveh, part of the royal archive from the most famous library in the ancient near East. The library was built by King Assurbanipal who reigned during Assyria's ascendancy in 8 BC. This historical treasure was completely destroyed when a combined force of two other races, the Medes and the Chaldeans, sacked Nineveh in 612 BC burying the library completely and robbing future generations of its royal archive. However, the flames thought to have destroyed the library are said to be responsible for firing the clay tablets into permanent records that lasted for centuries.

This unique collection includes tablets found in the ruins of the royal archive from the most famous library in the ancient near East. One set of seventy tablets from Nineveh revealed a vast programme of astronomical observations, which had been carried out in the second millennium BC during the Old Babylonian Period. Most of the tablets deal with interpretations of lunar and solar eclipses, conjunctions of planets and comets, which the Babylonians took as dangerous omens. Others are concerned with planets and the stars. These old records had been copied and stored at Nineveh in order that the local scribes would be able to understand future signals in the heavens.

What these tablets reveal is an unexpected link with an ancient scientific community dating back to before 1000 BC from which we can then establish a continuous link to modern cosmology. These early scientists were the first people whose observations and astronomical knowledge were accumulated over centuries in ancient Mesopotamia, eventually to be used by the Greek astronomer, geographer and mathematician, Hipparchus. Hipparchus' work is linked to the present day via Ptolemy, Copernicus, Kepler and Newton.

According to astral sources such as circular and tabular astrolabes and various observational reports to the kings in astronomical diaries, Babylonian astronomy goes back at least as far as 1800 BC. It appears to be focused mainly on the problem of establishing an accurate calendar, with an emphasis on recording and calculating the motions of the Sun and Moon.

Another motive for the compulsive observation of these two heavenly bodies did exist, namely the belief in the fateful meaning of certain alignments in the heavens, and in particular, solar eclipses. The extensive continuity of the Babylonian civilization enabled records to be kept over very long periods of time such that features like the gradual precession of the equinox and the regular cycles of solar eclipses could be recognized and studied. The Babylonians divided the sky into zones, the most important being those which lay along the ecliptic, the apparent path followed by the Sun, Moon and planets. These zones were limited to the area called the Zodiac. Latin names for the signs of the Zodiac as we know them today are translations of the old Babylonian constellations.

Babylonians appear to have been motivated by religious-philosophical reasons to take note only of isolated events. A planet's first and last appearances in the sky, for example, were noted rather than the systematic paths of the planets. Such occurrences were taken to have astrological significance because of the chance that they might foretell human fate. Though no extant evidence suggests that the Babylonians, unlike the Greeks, came up with any geometrical model of the cosmos, at the height of its creativity (around 600 BC) Babylonian astronomy could predict planetary motions with surprising accuracy.

The Babylonian Period

When Hammurabi, the Semitic king conquered the Sumerians, completing the unification of the region 'between the two rivers', Babylon became the capital city of his kingdom. Located on the left bank of the Euphrates, some 70 miles (113 kilometres) south of modern Baghdad, Babylon was ruled by the Hammurabi dynasty during what is referred to as the 'Old Babylonian Period' of 2000–1600 BC.

Following this 'Golden Age', when Babylon became the leading centre and capital, the whole region became known as Babylonia. Thus exists the convention of calling the mathematics and astronomy of this region 'Babylonian' even if they were not always originated or developed in the city of Babylon.

The rich heritage of literature, religion and astronomy from this period, found in the ruins of the ancient cities, would never have endured without the existence of a durable record-ing medium. The cuneiform clay tablets handed down from the Sumerians were perfect. These tablets were made from soft clay and written upon with the wedge-shaped stylus from which the name cuneiform is derived. The Latin word *cuneus* means 'wedge'.

A completed tablet was dried or baked until hard and usually protected by a clay case or envelope. Practically indestructible when dried, these tablets have given a wealth of information to modern scholars from this period, including thousands of astronomical and mathematical records. The ancient site of Nippur, once the site of an astronomical observatory in the Assyrian kingdom, for example, has alone produced 50,000 tablets.

The Old Babylonian period was a time of great advancement for the region. During this era most of the religious beliefs that developed encouraged the growth of a sophisticated astrology. Astrologer-priests made predictions of pending disasters based on celestial omens. By the beginning of the first millennium BC, the Babylonians had a highly developed writing tradition, sky-watching skills that had been applied in the creation of a calendar and a system of mathematics that was used to track the motion of the Sun and Moon.

Other studies of calendar making by ancient civiliza-tions such as the Egyptians and the Chinese show impressive schemes of constellation maps and Sun and Moon tracking, all designed to solve the problem of the synchronization of the Moon's motion with the Sun. Though the Moon provides a convenient time cycle for dividing the year, it has no bearing on the all-important seasons, which depend on the Sun. But the Babylonians went beyond others in their tenacity to use the Moon's cycle as a time keeping device. To do this they system-atically approached the Moon's motion in a matter not unlike the way natural science is carried out today.

Babylonian astronomers started with careful observations of the Moon's motion. These observations were accurately recorded over long periods of time. Next, they searched for repeating patterns in their records. Finally, they simulated these patterns using mathematical models to predict future positions. This may seem like a description of *modern* applied mathematical science, but is in fact how the Babylonians studied the motion of the Sun and the Moon during the first millennium BC.

Developing a lunar-solar calendar was relatively simple compared to their more ambitious goal of describing the complete movements of the primary heavenly bodies. By so doing, the scribes wished to anticipate as much as possible the occurrence of a lunar or solar eclipse, one of the most feared omens to appear in the sky. An eclipse of the Sun or the Moon was an awesome sight for the ancients. There is much evidence from early societies that they were shaken by the darkening and disappearance of the two celestial bodies which seemed to govern and sustain their existence. The sky was a dominating feature of that world, a fact since obscured by the prevalence of artificial lighting and different modes of time keeping. The regularity of celestial events provided order to early understanding of the cosmos.

Careful observations of the heavens allowed early stargazers to establish a division of time that enabled the development of calendars. Calendars, in turn, allowed for the planning of increasingly complex activities. Predicting the recurrence of the seasons (for agriculture) and reference points in the sky (for more extensive navigation) was essential in the development of a broader world view. An eclipse jeopardized this order and regularity.

By the third millennium BC, the Babylonians had become obsessed with celestial omens. Unlike the Egyptians, who had absolutely no interest in the dozens of eclipses that crossed the Nile during this same period, the Babylonians seemed so concerned about eclipses of the Sun and the Moon that they developed elaborate schemes to record these occultations over very long periods of time. This kind of record keeping suited them well. As J.J. Finkelstein of Berkeley has explained in his paper on Mesopotamian historiography (1963):

> To the Mesopotamian, the crucial and urgent study was the entire objective universe, without any interposition of the self between the observer and the observed. There probably has never been another civilization so single-mindedly bent on the accumulation of information, and on eschewing any generalization or enunciation of principles.

The Babylonians thus had compelling reasons for looking to the heavens. As the heavens were generally thought to be the home of the gods, Babylonians tried to read their destiny in unusual celestial happenings. For example, a letter from a diviner from the time of Hammurabi (about 1780 BC) reports on an eclipse of the Moon, which he suspected was a bad omen. During the same Hammurabi period a short manual of celestial omens appeared with the following instruction: 'If, on the day of its disappearance, the god Sin (the Moon) slows down in the sky (instead of disappearing suddenly), there will be drought and famine.'

Although celestial omens from the Old Babylonian period are known, more substantial development only came in the first millennium BC. Indeed, it was the thousands of artefacts of astronomical divinations that were found at the famous library at Nineveh that then produced the thousands of clay tablets referred to above.

In 536 BC, after seventy years of supremacy, the Babylonian empire came to an end when it fell to the Persians.

Assyrians: Warriors and Astrologers

The Assyrians were an extremely war-like people living around Assur in the Tigris valley in about 1100 BC. These people destroyed the first Babylonian state and extended their boundaries towards Asia Minor and Armenia. The new capital Nineveh was the political centre of a large military empire and as such was adorned with magnificent buildings made of the ubiquitous clay. Babylon, the great and rich commercial centre whose wealthy citizens largely governed themselves, retained its rank as a venerable seat of ancient culture. The Assyrian kings recognized the importance of Babylon at first, taking their oaths of office there, but in 689 BC the Assyrians turned against the great city and had it destroyed.

Yet they did not wipe out the Babylonian's fascination with stargazing. Having already adopted the ancient and quasi-religious practice of 'divination', they also absorbed the mathematical methods carried over from the Old Babylonian Period. Their rulers employed specialists in divination to continue the tradition of recording and interpreting eclipses and conjunctions of the Moon with planets, planetary movements, meteors and comets. A superstitious fear of calamity coupled with the belief in negotiating with the gods led to the intense interest in predicting eclipses. This in turn gave birth to the development of a programme of stargazing not unlike that practised by the Babylonians before them.

The Assyrians applied their skills of organization and discipline, building astronomical observatories with temple towers throughout the region. Over the period dating from 709–649 BC, reports were prepared which indicate not only detailed observations, but in the case of unfavourable eclipses, attempts at prediction. As the divination cult decreed, a successful prediction provided an opportunity to make supplication against any anticipated danger to come.

In time, however, the Assyrians were conquered by the Chaldeans, the last dynasty to rule in Babylon before Cyrus the Great's conquest by Persia. According to ancient historical writings, Persians were also known for their predictive skyscience and their obsession with celestial observations. There are conservative estimates that these people observed 373 solar eclipses and 832 lunar eclipses during their history, an impressive record given the rarity of this phenomenon.

As legend has it, Nabonassar destroyed all the records of

the previous kings of Babylon so that the reckoning of the Chaldean dynasty would begin with him. This new beginning was so effective that, centuries later, the Egyptian astronomer Ptolemy used Nabonassar's reign to fix the beginning of an era ('the reign of the Babylonian kings'). This was because he felt that the earliest usable observations began at this time. He went so far as to suggest that the era began at midday on 26 February 747 BC.

The date of 26 February 747 BC also marked an important beginning in the history of astronomy, because from this date, highly accurate astronomical observations by the chaldeans were kept on a regular basis until after the birth of Christ. Although the motive for these reports was still mainly astrological, these observations became increasingly what can only be described as scientific. Astronomical texts reveal that through centuries of pre-eminence under the Chaldean dynasties and later even during periods of decline, the celestial observations continued at Babylon on a regular basis with little change of pattern.

Modern scholars estimate the programme lasted almost eight hundred years. The most recent surviving astronomical text dates from AD 75, an almanac prepared from contemporary observations. Thus, from 750 BC to AD 75, the watch keepers at Babylon recorded what they saw in the heavens onto clay tablets. These tablets, which may be the most remarkable extant archive in history, are now stacked in the British Museum in the UK.

To give this achievement some perspective, consider an equivalent project to obtain similar observations at Windsor Castle starting at about the time of its construction in the early thirteenth century, during the time of Richard I and the Magna Carta. If the time depth of the Windsor 'archives' were to match Babylon's, sky watching would still be going on today – having continued through the reign of the Plantagenets and the War of the Roses, the marriage celebrations of Henry VIII, Elizabeth and the Spanish Armada, the Civil War, the Interregnum and the Restoration. Perhaps by the late seventeenth century observations would have been taken over by the Astronomer Royal and visited by Newton and Halley during the Glorious Revolution. During Queen Victoria's reign, her husband Albert, the overseer of great civic works, would no doubt have supervised the project. In the twentieth century, scribes would get deferments from the Great War, survive the blitz of the Luftwaffe and even the celebrations of the end of the Millennium.

The priests and scholars responsible for this remarkable programme of observations recorded continuously for over eight hundred years could be called 'Babylonian watch keepers'. As the centuries passed, mathematical models were applied to reproduce past observations and predict future movement of heavenly bodies, and the cult of astrology became more and more like what we now know as astronomy.

Though the greatest concentration of these observatories was in Babylon and the towns near it, the Assyrian records are the most complete as a result of the sacking of Assurbanipal's great library at Nineveh in 612 BC. This era was recognized by later historians as a turning point in the history of science. In the centuries to follow, increased accuracy in observations and the applications of mathematics turned the work of the scribes in Babylon into a science.

The Zodiac and the Celestial Sphere
With a strong mathematical tradition dating back to the Old Babylonian Period, Chaldean astronomers began to develop mathematical theory, relegating observations to a more a minor role. Analysis of the records of ancient observations suggests a model of mathematical simulation based on the celestial sphere shown in the figure overleaf. This model made the prediction of current and future astronomical phenomena possible.

However, more accuracy was demanded. As early as 1000 BC the scribes had recognized 18 constellations through which the Moon, the Sun and the planets always appeared to move. By 500 BC these constellations were systematized in such a way

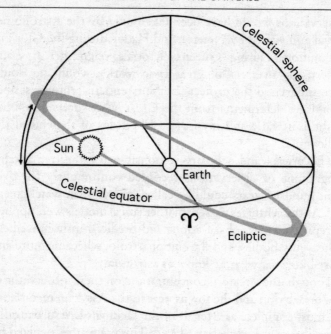

The Celestial Sphere Model.

that they were distributed among the twelve months, individually or sometimes in pairs. For example, the second month of the Babylonian year (which corresponds to mid-April to mid-May), had symbols of both Taurus and the Plaeides; the third month Gemini and Orion; and the twelfth month Pisces and Pegasus. This ring of constellations that lines the ecliptic was called the zodiac.

In an attempt to gain precise markings for the astronomical diaries and observations, the ecliptic path on which the Sun moved was divided into 12 equal parts of 30 degrees, making a total of 360 degrees, a complete rotation. This scheme was adopted after the Persians conquered Babylon in 538 BC but only used for the first time in a diary in 464 BC.

By about 400 BC the zodiac constellations had become the clearly defined zodiac that is used today. Beginning with

Constellations of the Zodiac on the ecliptic.

Aries (which corresponds to the time period of mid-March to mid-April), each covers 30 degrees of the sky. The zodiac constellations used on the celestial sphere model of the heavens to locate the Sun, Moon and planets is a scheme that has essentially lasted to the present day.

Up until the seventeenth century, the main motivation for the study of the movements of the heavenly bodies was the need of astrologers to have tables listing future positions of the Sun, Moon and planets. As we will see, some of the greatest figures in the history of astronomy had to pander to their patron's requests for better accuracy and improved astrological prognostications. For the Babylonians, this was a matter of life or death and the scribes learned to predict positions by using their sophisticated numerical system to take full advantage of the cycles revealed by their observational records. Columns of numbers were used to investigate special configurations of heavenly bodies such as repeating cycles of the alignments of the Earth, Moon and Sun.

The most important achievement of this entire early period from the standpoint of the history of science was the

Babylonian solution to the problem of the motion of the Sun and the Moon. It was their custom to designate the day after the new Moon as the beginning of each month, that is, when the lunar crescent first appeared after sunset. Originally, this day was determined by observation, but later they wanted to calculate it in advance – that is to say, to make a calendar.

By about 400 BC, astrologers realized that the Sun and the Moon's apparent motions around the ecliptic did not have a constant speed. These bodies appeared to move with increasing speed for half of each revolution to a definite maximum and then to decrease in speed to the former minimum. The Babylonians attempted to represent this cycle arithmetically by giving the Moon a fixed speed for its motion during one half of its cycle and a different fixed speed for the other half. Later they refined the mathematical method by representing the speed of the Moon as a factor that increases linearly from the minimum to maximum during half the revolution and then decreases to the minimum at the end of the cycle.

With these calculations of the lunar and solar months, Babylonian astronomers could predict the time of the new Moon and the day on which the month would end. They could accordingly predict the daily positions of Moon and Sun for every day during the month.

This mathematical science reached the height of its creativity during the Seleucid period. Predictions were essentially reduced to arithmetic, and though this led to highly precise predictions, the Babylonians never considered any geometrical models of the cosmos, (like the celestial sphere) which may have further supported their calculations. Problems were ultimately solved arithmetically without recourse to the cosmology that was to follow.

The Decline of Babylon

Inevitably, the Chaldeans did not maintain political power in Babylon. After less than one hundred years, their empire was overthrown by a powerful alliance of Medes and Persians in 538 BC. Babylon then became part of the Persian Empire under Cyrus the Great. Though the days of independent Mesopotamian kingdoms were over, astronomical observations still continued.

During more than two centuries of Persian rule, Babylonian astronomy continued to improve in the accuracy of both observation and mathematical predictions. The city of Babylon again shifted its political allegiance in 330 BC when the Persians were conquered by the armies of Alexander the Great. This began one of the most significant periods of cultural diffusion in all of history: the Hellenistic period.

Alexander planned to restore some of the glory of Babylon by making the city his eastern capital. He died there in the palace of Nabuchadrezzar II in 323 BC, and in the wars between his successors, Mesopotamia suffered much from the passage and the pillaging of armies. When Alexander's empire was divided in 321 BC, one of his generals, Seleucus, received the province of Babylonia to rule.

With the aid of Ptolemy, Seleucus was able to enter Babylon in 312 BC. He held Babylon against the forces of Antigonus for a short time before marching east, where he consolidated his power. It is uncertain when he returned to Babylonia and re-established his rule there; it may have been in 308, but certainly by 305 BC he had assumed the title of king.

With the defeat and death of Antigonus at the Battle of Ipsus in 301 BC, Seleucus became the ruler of a large empire stretching from modern Afghanistan to the Mediterranean Sea. He founded a number of cities, the most important of which were Seleucia, on the Tigris, and Antioch, on the Orontes River in Syria.

As a result, the population of Babylon was forced to move to the newly founded metropolis of Seleucid 62 miles (100 kilometres) north. The old city of Babylon never recovered from the removal of its intellectual and political core.

At Seleucid, a highly advanced form of Greek astronomy developed. The community made a particularly important advance by reviving Aristarchus of Samos' hypothesis that the sky could be explained by assuming the Earth turns on its axis once every 24 hours and along with the other planets revolves around the Sun. Most Greek philosophers who could not believe that the big, heavy Earth could revolve around light, celestial bodies, later rejected this explanation.

The Babylonian scribes had eventually learned to produce ephemerides, tables illustrating future positions of the Sun, Moon and planets. So in addition to the very careful record of observations that were available, the scribes made use of their numerical system and their knowledge of mathematics to take full advantage of the cycles revealed by their observational records. Once the ephemerides were completed, astrologers could make predictions without the observations and in all kinds of weather. This new mode of providing written, predictive information had hugely important implications to the future of astronomy, meteorology and navigation.

The marriage between astrology and astronomy has had a strange history. There is no question that during the Babylonian era the motivation for collecting celestial data was to satisfy the inclination for astrological knowledge. For many hundreds of years, particularly during the late first century BC, scribes and priests increasingly obtained this data.

The Babylonians were never attracted to the form of astrology popular today, that which is based on Greek geometric cosmology. Greek cosmology uses the celestial sphere model and the zodiac to interpret personality traits for horoscopes, with birth charts drawn from ancient myths and legends. With Babylonian astrology, celestial configurations were not the final word and emphasis was placed on the avoidance of unfavourable prognostications. This philosophy was therefore quite different than the version of astrology handed down to today's practitioners from the Greeks.

The difference between these two approaches was mainly due to the influence of two major writings from Ancient Greece, Plato's *Timaeus* (360 BC) and Ptolemy's *Tetrabiblos* (late second-century BC). Thereafter, classical astrology gave a naturalistic rationale for natal horoscopes, marking the split between astrology as 'divination' (Babylonian) and astrology as 'science' (Greek). The latter eventually split off and grew into what today is recognized as the science of astronomy. It is unfortunate that historians of science do not today interpret this split between astrology and astronomy in a positive way. This is undoubtedly due to the proliferation of untenable claims to interpret birth charts and horoscopes, which have since popularized the field of astrology, though in an unscientific way.

There can be little doubt that without the influence of astrology, astronomical observations from the ancient world would never have existed. Though many astronomers and other scientists loathe the concept, astrology has made an important contribution to the history of science. The thousands of dusty clay tablets in the British Museum, a product of early man's attempt to placate the gods, may hold yet more secrets of ancient celestial configurations. A major exhibition of Babylonian artefacts was mounted in 2008 at the British Museum in London and notably featured the importance of these artefacts in early pre-Greek Astronomy.

2

GREEK ASTRONOMY AND THE BIRTH OF NATURAL PHILOSOPHY

The next phase of the development of astronomy shifted dramatically from the priests of Mesopotamia to the philosophers of the Aegean. These inhabitants of the Mediterranean set the foundation for what was to become the science of astronomy as it was passed into Western Europe and then down to the present day.

There is a direct connection between the Babylonians and Hipparchus, the earliest of the great Greek astronomers, who proposed a geometric model for the non-uniform motion of the Sun based on Babylonian observations. Hipparchus lived from 190 BC to 120 BC, mostly on the island of Rhodes, and is only known by one surviving work and fragments from his lost star catalogue. Both of these documents later influenced the Alexandrian Claudius Ptolemy, who lived in the second century AD and helped develop the geometric model of the heavens that lasted until the time of Copernicus, fourteen centuries later.

Hipparchus' discussion of the extremely slow motion of the solstice and equinox points along the ecliptic from east to west against the background of the fixed stars is perhaps his most famous achievement. This is now referred to as the 'precession of the equinoxes'. This basic idea was also adopted – virtually unchanged – three hundred years later by Ptolemy.

It is clear that Babylonian astronomical records influenced Hipparchus, who made use of their methods as well as their observations to develop quantitative geometrical models, which describe the motion of the Sun. Though many questions remain unresolved regarding the relationship between Babylonian and Greek astronomy, Hipparchus' work provides a clear link between the two. Historians argue that he was responsible for the direct transmission of both Babylonian observations and methods and the successful synthesis of Babylonian and Greek astronomy.

Hipparchus was born in Nicaea, in what is now known as Iznik, Turkey and probably died on the island of Rhodes where he is known to have been a working astronomer from 147 BC to 127 BC. He is considered by many historians to be the greatest astronomer of antiquity and is the earliest Greek whose quantitative, accurate models for the motion of the Sun and Moon survive to the present. In order to achieve this, he had to make use of the observations and perhaps even the mathematical techniques accumulated over centuries by the Chaldeans from Babylonia. He thus demonstrated from the very beginning of his creative work in astronomy, that there is a fundamental importance in the interplay between observation and theory.

During his lifetime he formulated the world's first accurate star map, a catalogue of over 850 stars with estimates of their relative brightness. He also developed the system of epicycles that preserved the Earth-centred universe of Aristotle and supported the motion of all heavenly bodies in perfect circles. This system was in approximate agreement with the observations of the time, almost three hundred years before Ptolemy, who is introduced later in this book. Perhaps what is more surprising to any student of the history of astronomy is that Hipparchus also produced the first eccentric orbital construction to predict the motion of the Sun. This simple construction was used almost unaltered (but acknowledged) by Ptolemy three centuries later as part of the elaborate model published in *The Almagest*, which was then handed down to Copernicus in the sixteenth century.

In Hipparchus' simple geometry the Sun orbited the Earth on a circle, moving about the centre of the circle at a uniform speed (and therefore satisfying Aristotle's conditions). But Hipparchus knew unquestionably, probably from his access to Babylonian records, that seasons are different lengths and that this must be accounted for in any orbital model. Clearly, the Earth could not be located in the centre of the Sun's orbit, but must be eccentric (i.e. off centre) in order to reproduce its non-uniform motion. He was able to calculate just how far the Earth had to be displaced from the centre of the Sun's orbit in order to give agreement with the speeding up and slowing down of the Sun during its journey around the ecliptic – the path of the Sun as seen from the Earth on the background of the stars. This single off-centre eccentric was sufficient to reproduce the motion of the Sun throughout the year.

Incorporated directly into Ptolemy's model, the eccentric later became quite controversial as Copernicus and Kepler began undermining geocentrism, as we will see in subsequent chapters. The Moon has a more complicated motion than the Sun. It is in Hipparchus' model for the Moon's motion, that we begin to more clearly recognize the shift to geometric astronomy that the Greeks began through their assimilation of the Babylonian commitment to observations. Hipparchus used astronomical records and parameters derived from Babylonian sources in his development of quantitative geometric models for the motion, of both the Sun and the Moon. In so doing he ultimately demonstrated the fundamental importance of the interplay of observation and theory for the future development of astronomy

Hipparchus' Star Catalogue

Study of the Babylonian records produced other rewards for Hipparchus. As a systematic observer himself and one of the few Greek astronomers who valued the early records of the Babylonians, he was able to compile a remarkable catalogue of stars. By combining his own observations with the older, more extensive measurements from Mesopotamia, he was able to observe the movement of any one star over the centuries. He continued to pursue the historical plotting of stars and was eventually able to distinguish the slow movement of the entire sky, including the drifting of the equinoctial points, the astronomer's main reference frame, from east to west among stars. Equinoctial points are the places where the celestial equator is crossed by the path of the Sun, i.e. the ecliptic. The Vernal, or spring equinox, is the point at which the Sun's path crosses the celestial equator in the ascending direction. This precise crossing had, for centuries, been understood by astronomers to mark the beginning of the celestial year. On the autumn equinox, the Sun crosses in the descending direction.

Hipparchus' discovery, that this particular point is steadily moving, indicated that the measured position of a star varies with the date of the measurement. This is called the precession of the equinoxes and is now known to be approximately 1 degree every 70 years. To detect such a small change without a telescope was quite an achievement for an observer in the second century BC.

This was Hipparchus' single most important discovery. Knowing as he did that very long-term observations would be necessary to confirm this phenomenon, he recorded the positions of hundreds of stars. This enabled successive astronomers to measure and compare the future movements of stars. In the process Hipparchus compiled the first comprehensive catalogue of stars in the western world. His star catalogue has been since lost, although a few partial star positions are recorded in his only surviving work, the *Commentary* (c.276 BC). In addition, Professor Bradley E. Schaefer of Louisiana State University in Baton Rouge, Louisiana has found evidence of an image of Hipparchus' star catalogue on a marble celestial sphere, part of an extant ancient Roman sculpture now in the Farnese collection of the National Archaeological Museum in Naples.

The Greek Philosophers
Though Hipparchus may with some accuracy be called Greece's first astronomer, he was not the first philosopher from the Aegean

to contemplate the heavens. Nearly all historians give the ancient Greeks credit for introducing the idea that numerical relationships can manifest themselves in the physical world. As we have seen, they did so by using arithmetic and geometry combined with logic to provide an explanation for astronomical events. On another level, Greek metaphysics implored humans to be curious, to seek truth, to look for patterns and to use reason to solve problems. These ideals remain the central tenets of science and scientific exploration today, in the twenty-first century.

Perhaps the first man to call himself a philosopher was Pythagoras, who was born on the island of Samos, close to the mainland of what is now known as Turkey. Pythagoras made influential contributions to philosophy and religious teaching in the early sixth century BC. He also exercised a marked influence on another important Greek philosopher, Plato. Pythagoras founded a religious movement based on the belief that everything was related to mathematics and that numbers represent the ultimate reality.

Pythagoras taught that all observable phenomena could be measured and predicted based on rhythmic patterns or cycles. Unfortunately, very little is known about the man himself because none of his writings have survived. Some historians believe that his colleagues and successors may actually have made many of the accomplishments credited to Pythagoras. Nevertheless, his name has come to symbolize the order, harmony and simplicity of numbers that have applied by subsequent generations to the structure of the universe.

We can be more certain about the existence of the next giant of classical philosophy. Socrates lived and taught in Athens during the Golden Age of Greece (circa 546–404 BC). The age began with the unlikely victory of a badly outnumbered Greek army over a vast Persian army. Following this victory, significant advances were made in a number of fields, including the formulation of the structure of democracy, art, philosophy, drama and literature. Some of the Greek names most familiar to us, such as the renowned military and political leader

Pericles, lived in this exciting and productive time. It was an era marked by such high and diverse levels of achievement that classical scholars coined the term 'the Greek miracle'.

Although Socrates did not leave any written work, he is included in this listing because his ideas were documented by his student Plato and later transmitted to Aristotle, the man who first documented a theory of science. Plato witnessed the notorious death of Socrates, a result of being charged with 'corrupting the youth of Athens'. In spite of this, Socrates' views on morality and his disdain for the physical world set the stage for Plato and his prize student Aristotle to build wide reaching and coherent world views. These later provided the foundation for much of western thought.

Early in the fourth century BC, Plato adopted an intellectual approach to the study of celestial motion. He defined the problem to his students in the Agora in Athens. It is with Plato that we find the birth of the notion of the 'perfection of the heavens' that was to hinder astronomical thinking for two millennia.

In terms of quality and importance, Plato contrasted bodies in the heavens with objects on the Earth. The stars, he said, represent eternal and divine unchanging things that move with uniform speed around the Earth in the most regular and perfect of paths, the circle. This idea took root even though it was well known that while the motion of the Sun and Moon seemed regular, the planets – which are also celestial objects – wandered across the sky in complex paths, occasionally veering back before moving forward again in their orbits.

Plato contended that if the motions of heavenly bodies do not move in a perfect circle then the movement must be described by some combination of perfect circles. This planted the idea for the Greeks that the motion of all heavenly objects could only be described by perfect circles and uniform speed. And thus Plato's devastating legacy took root. This principle, which was clearly inspired by the strong influence of Pythagoras, became a major restriction on the thinking of generations of philosophers and scientists in Greece and beyond.

Aristotle

Not long after Plato's pronouncements, the principle of uniform circular motion in the heavens was further ingrained into Greek philosophy. This was achieved as a result of an elaborate scheme for defining the natural world pioneered by Plato's most famous student, Aristotle. The legendary Athenian – soon to become the master of Greek thinkers and famous as the teacher of Alexander the Great – set forth a detailed list of the 'natural conditions of things' which included Plato's heavenly principle as well as many other ideas on what we would today call 'science'. Much of the subsequent thinking of Greek philosophers on nature would follow from his ideas.

Here is a summary of his postulates:

* There is order in the universe: the baser elements lie at the bottom and the nobler elements at the top.
* The Earth is the basest of all objects in the universe, therefore it is at the bottom.
* The Earth is composed of four elements: air, water, fire and earth. These elements always seek each other. Thus, air and fire rise and water and earth fall.
* Once the materials have regained their rightful place, their natural tendency is to remain motionless.
* Sideways motion is caused by violence (or force). Eventually this violent motion runs out and the object either rises or falls according to its nature.
* Violent and natural motion affects only the four elements.
* Celestial bodies are made of a fifth element called *quintessence* and their natural behaviour is to move uniformly in circles indefinitely around the Earth (which does not move) either by spinning or by moving through the heavens.
* The heavens are eternal and unchanging, whereas the Earth is subject to decay and change.

Most students today can easily recognize the flaws in Aristotle's 'science' and it might seem strange that these ideas found acceptance amongst both his contemporaries and the many later generations who read and accepted his works as absolute wisdom. To understand this it is important to remember that Aristotle's concepts of astronomy and physics were intertwined with his ideas on philosophy and logic as well as his concepts of social and political order. The important concept of geocentrism, the Earth at the centre of all things, therefore slipped into Aristotle's cosmology without much alarm. If to Plato it seemed self-evident that heavenly bodies move uniformly in perfect circles, to Aristotle and his contemporaries, it was self-evident that the Earth was stationary at the centre of these circles.

Because Aristotle's cosmology has had a long lasting – albeit regressive – effect on the progress towards our modern view of astronomy, it is helpful to review its details. A great philosopher he may have been, but as a physicist or cosmologist he relied too much on an a *priori* approach to the natural world, relying on intuition rather than insisting on testing nature with observations and experiments. Some historians of science would say that although Aristotle may have been wrong, he was not 'unscientific' and should still be credited with the birth of science. This view may have some validity if one considers Aristotle's starting point. However, his view is ultimately flawed given that the observational verification of theory is so fundamental to what we call 'science' today. Given that the history of science demonstrates many examples of phenomena in the physical world contradicting intuitive learning or 'common sense', the necessity for experimentation is now generally accepted.

From its inception around 350 BC, Aristotle's cosmological work *On The Heavens* was the most influential treatise of its kind, accepted as truth for more than eighteen centuries. In this work, Aristotle discussed the general nature of the cosmos and the physical properties of individual bodies, essentially defining the field of 'science' for the first time.

Since Aristotle had postulated that all bodies are made up of four elements: earth, water, air and fire; he further theorized that composite objects have the features of the dominant of these elements in its composition. Most things are of this mixed variety since few objects on Earth are totally pure substances. From this, Aristotle concluded that things on the Earth are imperfect.

The natural tendency of objects was all-important to Aristotle and his idea that all bodies, by their very nature, have a natural way of moving is central to Aristotelian cosmology. Movement is not, he states, the result of the influence of one body on another (as we understand it today from Galileo and Newton) but the result of the simple premise that bodies have natural movement: some naturally move in straight lines, some naturally move in circles and others naturally stay put. We know today that the natural state of a body's motion is at uniform speed in a straight line unless acted upon by an external force. This is called the principle of inertia, which was not discovered until the careful experiments of Galileo twenty centuries after Aristotle.

The most important of the natural movements that Aristotle identified is the circular motion. Since he assumed that for each natural motion there must be a particular corresponding body, Aristotle presumed that heavenly bodies move naturally in perfect circles, as Plato had taught. He accordingly then postulated that heavenly bodies are made of a more perfect substance than earthly objects. Finally, as a last a *priori* conclusion about heavenly bodies, Aristotle stated that since the stars and planets are so special and move in circles, it is also natural for these objects to be perfectly spherical in shape.

Another of the fundamental propositions of Aristotelian philosophy is that there is no effect without a cause. Applied to moving bodies, this proposition means that there is no motion without a force. Speed, then, is proportional to force and inversely proportional to resistance. This notion is not at all unreasonable if one takes an ox pulling a cart as the defining case of motion: the cart only moves if the ox pulls, and when the ox stops pulling, the cart stops.

When Aristotle applied his rule to falling bodies, he found that the force was equal to the weight pulling a body down and the resistance is that of the medium (air or water, for example) through which the body falls. When a falling object gains speed, Aristotle attributed this to a gain in weight. So, if weight determines the speed of fall, then when two different weights are dropped from a high place, the heavier will fall faster than the lighter, in proportion to the two weights. A 10-pound weight would reach the Earth by the time a 1-pound weight had fallen one-tenth as far. This concept was yet to be refuted when, according to the well-known but probably apocryphal story, Galileo dropped different weights from the Tower of Pisa in the sixteenth century.

If we were to accept Aristotle's theory we must consider the cosmos to be made up of a central Earth (accepted as spherical) surrounded by a Sun, Moon and stars all moving uniformly in circles around it – a conglomerate he called 'the world'. Note the strange idea that all celestial bodies are perfect, yet they circle the imperfect earth. Going further, Aristotle hypothesizes that the initial motion of these spheres was caused by the action of a 'prime mover' who acts on the outermost sphere of the fixed stars, with the motion then trickling down to the other spheres. Though one might argue that the great man's logic was sound, Aristotle was wrong in many of his initial assumptions. Particularly misleading is the thesis that different objects have different natural motions.

In his description of the heavens, Aristotle created a complex system containing 55 spheres – an elaboration on the original sphere of an earlier astronomer Eudoxus of Cnidus. This system had the virtue of explaining and predicting most of the observed motions of the stars and planets and all of the characteristics of a scientific theory. He painstakingly modified the model, matching it to the observations available, until all these could be accurately explained.

Yet he did not consider the model a work-in-progress subject to continuous testing and experimentation, as is the norm today. He simply wished to use the model to make predictions, such as the position of a planet a year into the future, satisfying the Greeks' compelling goal to 'save the appearances'. In spite of these imperfections, this was the start of the development of the celestial sphere model later developed by Ptolemy, which came to be accepted and utilized by generations of astronomers and astrologers after Aristotle.

Alexander the Great

Aristotle inspired an avid hunger for knowledge among the Greeks. Against this background his pupil Alexander the Great launched his global enterprise of conquest in 334 BC. This was accomplished with meteoric speed until Alexander's untimely death about a decade later in 323 BC at the age of thirty-three. This was a year before Aristotle, who had lived twice as long.

Alexander's aim was not restricted to conquests, but also included explorations. He dispatched close companions – generals as well as scholars – to report to him in detail on regions previously unmapped and uncharted. His campaigns therefore resulted in a considerable expansion of empirical knowledge of geography. The reports he acquired stimulated and motivated an unprecedented interest in scientific research and the study of the natural qualities and inhabitants of the Earth. His Age was charged with a new spirit of enquiring.

Scholars have long seen Alexander's conquest of the Persian Empire as opening the floodgates for the spread of Greek culture in and around the Mediterranean. He attempted to create a unified ruling class of Persians and Greeks, bound by marriage ties, and used both in positions of power. He tried to mix the two cultures, encouraging intermarriage, adopting elements of the Persian court and also attempting to insist on certain Persian practices. Alexander also unified the army, placing Persian soldiers into the Greek ranks.

After Alexander's death, the Empire was split into separate states under his generals and although the kings who succeeded him rejected most of Alexander's cultural changes, other less definite policies were continued. The founding of cities was a major part of the struggle for control of any particular region, and the independence of Greek cities was a political right that his successors sought. They used the existing systems of government within their individual states, but often placed Greeks in the top levels of power. Not surprisingly, the spread of the Greek language also increased, and was often used in tandem with the native language for administrative purposes.

Four main kingdoms maintained Macedonian and Greek rule over native populations, and while they allowed the flourishing of native culture and religion, this was ultimately mixed with their own Greek culture.

The Library of Alexandria

On a visit to Egypt in 330 BC Alexander founded Alexandria, one of the many cities that were to bear his name. Leaving his administrator Cleomenes to build the new city as he made further conquests, Alexandria was destined to become the most important city in his new realm. However soon after Alexander's death, his empire was subjected to the greedy land-grabbing ambitions of his many generals.

Ptolemy I (305–282 BC), also called Soter, took Egypt and after disposing of Cleomenes, made Alexandria his capital. He and his descendants ruled from Alexandria for three centuries and made little effort to integrate the rest of Egypt into their Greek culture. The Ptolemys formed their coastal capital into the great intellectual and cultural centre of its age, immortalized by the magnificent library founded by Ptolemy I at the beginning of the third century BC. Its collection of papyrus scrolls – said to have numbered nearly half a million – attracted intellectuals from all over the Greek-speaking Mediterranean.

According to the earliest source of information – a second century BC letter of Aristeas paraphrased by the Hebrew historian Josephus – the Library was initially organized by Demetrius of Phaleron, a student of Aristotle, and was modelled after the arrangement of Aristotle's school in Athens. The library was closely linked to a museum built by the Ptolemys that seems to have focused primarily on editing texts. Initially, libraries were important for textual research in the ancient world, since the same text often existed in several different versions of varying quality and veracity. Later, the library took on a more important function as a repository for all the great books of the day.

When the founder Ptolemy I asked, 'How many scrolls do we have?' Demetrius was on hand to answer with the latest count. After all, it was Demetrius who suggested setting up a universal library to hold copies of all the books in the world. Ptolemy I and his successors wanted to understand the people under their rule and store Latin, Persian, Hebrew and Egyptian books, all of which were translated into Greek. The library's lofty goal was to collect half a million scrolls and the Ptolemys took serious steps to accomplish this ambitious feat. Ptolemy I, for example, composed a letter to all the sovereigns and governors he knew, imploring them 'not to hesitate to send him' works by authors of every kind.

The problem of growing the collection was solved through an innovative piece of legislation by Ptolemy III (246–222 BC), the third ruler of the Ptolemaic dynasty. He decreed that all visitors to the city were required to surrender all books and scrolls as well as any form of written media in any language in their possession. Official scribes then swiftly copied these writings and the reproduction was so precise that the originals were often put into the library and the copies were delivered to their unsuspecting owners. This process helped to create a substantial reservoir of books in the relatively new city.

The Ptolemys engaged in further acquisitions, some in an orthodox way, like purchasing writings from throughout the Mediterranean area; and some unorthodox, like confiscating any book not already in the library from passengers arriving in Alexandria. They were obsessed with becoming the most important library in the Hellenistic world and carried out some shocking tactics to achieve this goal. Ptolemy III, for example, deceived Athenian authorities when they let him borrow original manuscripts of Aeschylus, Sophocles and Euripides using silver as collateral. He kept the originals and sent the copies back, letting the authorities keep the silver.

Physically the books were not what we think of today, but rather scrolls, mostly made of papyrus, but sometimes of leather. They were kept in pigeonholes with titles written on wooden tags hung from their outer ends. Older copies were favoured – the older the better – since these would be considered more trustworthy. At its height, the library held nearly 750,000 scrolls, with works by Euclid, Aristarchus, Eratosthenes and Hipparchus. Amongst the most important documents in this vast collection were the writings of Aristotle, which contained his model of the heavens.

Euclid was a Greek mathematician of the Hellenistic period who thrived in Alexandria, almost certainly during the reign of Ptolemy I. His *Elements* is the most successful textbook in the history of mathematics. In it, the principles of Euclidean geometry are deduced from a small set of axioms. Euclid's method of proving mathematical theorems by logical deduction from accepted principles remains the backbone of all mathematics. He also wrote works on perspective, conic sections, spherical geometry and quadric surfaces.

Although many of the results in *Elements* originated with earlier mathematicians, one of Euclid's accomplishments was to present them in a single, logically coherent framework, making them easy to use and easy to reference. This includes the system of rigorous mathematical proofs that remains the basis of mathematics some twenty-three centuries later. Although best known for its geometric results the *Elements* also includes number theory and considers the connection between perfect numbers and primes – the infinitude of prime numbers – Euclid's dilemma on factorization and the Euclidean algorithm for finding the greatest common divisor of two numbers.

The geometrical system described in the *Elements* was long known simply as 'geometry', and was considered to be the only geometry possible. Today, however, that system is often referred to as Euclidean geometry to distinguish it from other so-called non-Euclidean geometries that mathematicians discovered in the nineteenth century. These include Riemann geometry, which was employed by Einstein in determining his general theory of relativity in the early part of the twentieth century.

An Early Heliocentric System

Aristarchus is an important figure in the history of astronomy even though his advanced ideas on the movement of the Earth were not incorporated into the development of the classic Greek model published by Ptolemy in the second century. He was certainly both a mathematician and astronomer and is celebrated as the first to propose a Sun-centred universe. He also is known for his pioneering attempt to determine the sizes of the Sun and Moon and their distances from the Earth. This idea survives in part because of the work of Archimedes and Plutarch, which further expands on Aristarchus' only extant work, a short treatise called 'On the Sizes and Distances of the Sun and Moon'.

In this notable work, Aristarchus provides the details of his remarkable geometric argument. Based on observation, he determined that the Sun was about 20 times as distant from the Earth as the Moon, and 20 times the Moon's size. Both these estimates were an order of magnitude too small, but the fault was in Aristarchus' lack of accurate instrumentation rather than in his method of reasoning.

It is due to the prestige of the great Archimedes that we are certain of Aristarchus' advanced hypothesis on heliocentrism, in which the Sun and not the Earth is at the centre of all things. In his well-known book, *The Sand Reckoner*, which he addressed to his patron King Gelon in ancient Syracuse, Archimedes describes how to count the number of grains of sand in the universe. In passing, he mentions the latest ideas about the universe from the mainland, reporting on the innovative hypotheses of Aristarchus:

> . . . His hypotheses are that the fixed stars and the Sun remain unmoved, that the Earth revolves about the Sun on the circumference of a circle, the Sun lying in the middle of the

orbit, and that the sphere of fixed stars, situated about the same centre as the Sun, is so great that the circle in which he supposes the Earth to revolve bears such a proportion to the distance of the fixed stars as the centre of the sphere bears to its surface.

Aristarchus' heliocentric system was indeed revolutionary. It was the Earth that rotates, he said, once daily on an axis of its own which causes the apparent daily motion of the stars. He believed that this assumption could explain all the daily motions observed in the sky. The observed angle of the paths of the Sun, Moon and the planets with the celestial equator results from the tilt of the Earth's own axis. Annual changes in the sky, including retrograde motion of planets, were then explained by assuming that the Earth and the planets revolve around the Sun.

In this model the previously assigned motion of the Sun around the Earth was now subverted so that the Earth moved around the Sun. The Earth essentially became just one among several planets. These bodies were then not made of 'heavenly material' to house the gods but were considered to be composed of material rather like that which we think composes the Earth today. There is much debate by historians about why the Greeks rejected the elegant and innovative idea of a Sun-centred universe.

The diagram shows how much Aristarchus' heliocentric system could explain retrograde motion, that is reversing the orbital motion of Mars, Jupiter and Saturn, the planets outside the orbit of the Earth. The outer planets, assumed to be moving around the Sun in circular orbits, move more slowly than the Earth and when the Earth passes directly between the Sun and the planet, an illusion is observed. To viewers on Earth, the outer planet appears for a time to move backwards in retrograde motion against the fixed background of the stars. The heliocentric hypothesis has one further advantage. It explained the observation that the planets were brighter during retrograde motion as the outer planets were nearer to the Earth during the passing phase.

Even with these consistent explanations, Aristarchus'

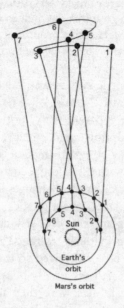

Explanation of retrograde motion in an heliocentric system.

proposal was severely criticized by his contemporaries for several basic reasons. First, the idea of a moving Earth contradicted the philosophical doctrines that the Earth is different from the celestial bodies and that its natural place is in the centre of the universe as proposed by Aristotle. Second, a proposal that the Earth moves was considered to be somewhat immoral, given its connection to the realm of the Creator. Thirdly, this new picture of the solar system contradicted common sense and the common observation that the Earth certainly seems to be at rest, not rushing through space.

Finally, the expected displacement of the stars as a result of the semi-annual journey of the Earth from one side of its orbit to another was not observed. This is shown in the figure on page 36 and is called parallax. The Greek astronomers never observed the parallax shift in the position of a star. Clearly, the drawing is exaggerated.

The null parallax result could be explained either because the Earth does not revolve around the Sun and therefore no shift will occur, or because the stars are so far away that the shift is too small to observe. The Greeks realized that if the displacement is really too small to detect, the stars must be an incredible distance from the Earth. Given that the scale of the universe generally accepted at that time was only a fraction of its true size, this argument gathered little support.

The parallax effect is in fact there, but it is very small because the stars are so far away that their parallax can only be observed with very powerful telescopes. Thus, the heliocentric idea of Aristarchus was forgotten and western thought about heliocentrism stagnated for almost two thousand years. Indeed, the parallax of stars was not measured conclusively until the year 1838, when the Prussian Friedrich Wilhelm Bessel used a high-powered telescope.

Perhaps the most compelling reason for the rejection of the heliocentric proposal was that this scheme required such a drastic change in people's image of the universe that Aristarchus' hypothesis had no influence on Greek astronomy. Furthermore, there is little evidence that this prescient work applied to the solar system had any influence on the thoughts of Copernicus in 1543. He made similar arguments – with similar criticisms – some eighteen centuries later, quite independently.

Finally, it is left to introduce Eratosthenes, one of the brilliant alumni of the Alexandria Library. Born around 276 BC at a Greek colony in Cyrene, Libya, he was educated at the academies of Athens and appointed to run the Great Library at the age of thirty-six. It is with Eratosthenes that much of what is now described as scientific scholarship began. The funds from the royal treasury that paid the chief librarian of the Alexandria Library and his scholarly staff provide the first example of scientific and literary grants. Commonly called the Father of Geography, Eratosthenes was the first to use the word 'geography'. His studies of the planet led him to determine the circumference of the Earth.

Having heard of a deep well at Syene in Upper Egypt (near modern Aswan) where sunlight only struck the bottom of the

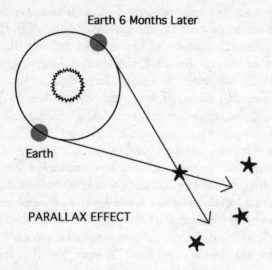

Measurement of parallax in heliocentric solar system.

well at noon on the summer solstice, Eratosthenes determined
from this that he could estimate the circumference of the Earth.
He therefore measured the angle of the shadow cast by a vertical
pointer at Alexandria at that same instant (noon on the summer
solstice). He noted that the shadow made an angle of 1/50th of
a circle. Thinking that the two points were separated by 5,000
stades (a stade is an ancient unit of length equivalent approximately to a modern measurement of 607 to 738 feet (about 185
to 225 metres)), he used a simple trigonometric argument to
calculate that the full circumference of the Earth was 50 times
5,000 stades, or 250,000 stades – amazingly close to the modern
measurement of 25,000 miles (40,000 kilometres).

The Coming of the Romans
Alexander and Aristotle died within a year of each other, and
with their deaths came the threat to Hellenism. Alexander
had tried to perpetuate the Greek culture throughout the

Mediterranean but Rome, posing a threat in the southern part of the Italy and on Sicily, also flexed its imperial muscles in Magna Graecia (Greater Greece). With the advent of the Pyrrhic War (280–275 BC) – a complex series of battles between the Greeks and the Romans – the rise of Rome became inevitable.

The Italian peninsula had been undergoing a gradual consolidation under Roman hegemony for centuries, and Rome's victory over Pyrrhus was an especially significant one. It was the victory for Rome over a Greek army that fought in the tradition of Alexander the Great and was commanded by the most able commander of its time who gave his name to the battle.

After the defeat of Pyrrhus, Rome was recognized as a major power in the Mediterranean. This was marked by the opening of a permanent embassy of amity by the Macedonian king of Egypt in Rome in 273 BC. New Roman colonies were founded in the south to further secure Roman domination, whilst in the north, new colonies were founded to cement Roman rule. Rome was now mistress of the entire peninsula: from the Straits of Messina to the Apennine frontier along the Arno and the Rubicon rivers.

It was only at the beginning of the second century BC, after a century of fighting, that Rome intervened directly in the Orient. They began with conquests of the Antigonid dynasty and Antiochus III the Great, the last great political figure of the Hellenistic sovereigns.

In a slow and complex process that lasted almost two centuries, Rome extended its rule in the eastern Mediterranean. The Roman General Pompey eliminated the Seleucid empire (which today includes parts of Pakistan and Turkmenistan) in 63 BC and reorganized its newly acquired eastern territory under Roman laws. This set the stage for the famous victory of the first emperor of the Roman empire, Augustus, over General Mark Antony, the infamous lover and ally of the last queen of Egypt, Cleopatra VII. This fight proved to be the last act of this conquest, as with Cleopatra's defeat and suicide in 30 BC, Roman penetration into the eastern Mediterranean was complete.

3

PTOLEMY: THE SHAPE
OF THE UNIVERSE

The burning of the Library during the Roman occupation of Alexandria under Julius Caesar, though possibly apocryphal, has been described as the greatest calamity of the ancient world. The most complete collection of all Greek and near eastern literature was lost in one great conflagration. Yet in reality, the Library and its community of scholars did not really disappear. The collection and the society it inspired survived through the Roman Empire and the incessant turbulence of the Empire's most volatile and valuable city.

Alexandria was a cosmopolitan city, and drew Greeks, Egyptians and Romans into a unique and not entirely harmonious coexistence. The Alexandrian Library and Museum had become an ideal place for scholars from these different cultures to meet and exchange ideas. What is more, it served as a repository for the literature and accounts of the Alexandrian intelligentsia and Roman Empire in general.

It seems probable that the Library and its facilities were still intact as a great centre of Hellenistic civilization when Claudius Ptolemy appeared on the scene in about AD 125. Ptolemy was to compile the successful elements of 500 years' worth of Greek geometric astronomy into a single system. This system was then adopted as the standard model and was used for over fourteen centuries.

Ptolemy's origins are shrouded in mystery. As previously mentioned, there was a general in Alexander's army who made himself King of Egypt in 323 BC, Ptolemy I Soter. All the kings after him, until Egypt became a Roman province in 30 BC, were also Ptolemys. However, Ptolemy the astronomer was not related to this royal dynasty. Nevertheless, he has been confused with the dynasty and, as a result, an iconographic tradition developed lasting a thousand years in which Ptolemy was always represented wearing a crown. Despite being a Roman citizen, scholars have concluded that from an ethnic perspective, Ptolemy was a Greek.

Beyond his being a member of Alexandria's Greek society during the Roman period, few details of Ptolemy's life are known. He wrote in ancient Greek and is known to have used Babylonian astronomical data and important parts of the work of Hipparchus. In fact, Ptolemy synthesized and extended Hipparchus' solar/lunar system of epicycles and eccentric circles as part of his own geocentric theory of the solar system.

By AD 150 in Alexandria, Ptolemy was poised to take advantage of the many important contributions from previous Greek astronomers that had accumulated in the great library. He knew he could build on many of his predecessors' ideas in developing his own astronomical system. With the usual motivation of an astrologer, he wished to be able to predict the positions of the known planets as well as the Sun and the Moon at any time in the future. He defines the problem at the beginning of his masterwork, *The Almagest* (dating from about AD 150) clearly indicating the assumptions he had adopted as his premise:

We wish to find the evident and certain appearances from the observations of the ancients and our own, and applying the consequences of these conceptions by means of geometrical demonstrations. We have to state that the heavens are spherical and move spherically; that the Earth, is sensibly spherical; and in position, lies right in the middle of the heavens, like a geometrical centre; in magnitude and distance, the Earth has the ratio of a point with respect to the sphere of the fixed stars, having itself no local motion at all.

Ptolemy then argued that each of the assumptions cited in the passage above was necessary and, what is more, was consistent with all available observations. He continued to cite evidence, for example, that the world was geocentric: 'It is once and for all clear from the very appearances that the Earth is in the middle of the world and all weights move towards it.'

His writing further supported his interpretation of astronomical observations by citing the physics of falling bodies in the manner of Aristotle. Thus, according to Ptolemy, the observations made by Hipparchus, himself and others were entirely in agreement with the elaborate proposals of the great Aristotle. Though he was aware of the heliocentric ideas espoused by Aristarchus, he gave little credence to these. In fact, he used his mixture of physics and astronomy to disprove the idea that the Earth might rotate and revolve. In a direct attack on Aristarchus, he wrote the following:

Some people [by which Ptolemy meant Aristarchus and his followers] agree on something which they think is more plausible. That there is nothing against supposing the heavens immobile and the Earth as turning on the same axis [as the stars] from West to East, nearly one revolution per day. But it has escaped their notice that as far as the appearances of the stars are concerned, nothing would perhaps keep things from being in accordance with this simpler conjecture [that is, Ptolemy's system] but that in the light of what happens around us in the air such a notion would seem altogether absurd.

He later wrote about the unusual phenomena that might occur if the Earth was to rotate, as heliocentric supporters believed:

If the Earth did rotate, it would not pull its blanket in the air around with it. As a result, all clouds would fly past towards the west and all birds and other things in the air would also be carried away to the west. Even if the Earth did drag the air along with it, objects in the air would still tend to be left behind by the Earth and air altogether.

Before introducing the details of Ptolemy's system, it is important to review what was known in the mid-second century. Quite a lot was known about the motions of the heavens by the time of Ptolemy and it would be naïve to think, in spite of the detailed description of his cosmology presented earlier, that Greek astronomy began with Aristotle. It is known, for example, that as early as 380 BC the Greek philosopher Plato (notably, Aristotle's teacher) recognized that the phases of the Moon could be explained by thinking of the Moon as a globe reflecting sunlight and moving around the Earth in about 29 days.

It is easy to see that the Moon's path around this sky is close to the annual path of the Sun. Another way of understanding this is that the Moon is always near the ecliptic, or the imaginary line on the celestial sphere which follows the path of the Sun. But the Moon's path is slightly tilted with respect to this line. If not, there would be a solar eclipse every month at new Moon when the Moon would be exactly between the Sun and the Earth. Likewise, at every full Moon, the Moon would be eclipsed as the Earth blocked the light from the Sun. As we know this does not happen, and the path of the Moon is actually known to be tilted with respect to the path of the Sun, that is, the ecliptic.

Even without a telescope, the Greeks could see at least five bright objects that moved among the stars, in addition to the Sun and Moon. This would have been particularly true in ancient times with no pollution to cloud the sky and no external lights causing glare. So, the Greeks could detect the motion of the following planets: Mercury, Venus, Mars, Jupiter and Saturn. The other planets, Uranus, Neptune and Pluto, would only be discovered after the invention of the telescope.

The planets generally move slowly eastward among the stars. But they have another remarkable and puzzling motion of their own which was ignored by Aristotle in his treatise *On the Heavens*. At certain times, each planet stops moving eastward among the stars and for some months moves back westward. This westward movement is called retrograde motion and was overlooked by Aristotle, who assumed that all heavenly objects moved uniformly across the sky, always in the same direction.

In order to explain retrograde motion, new schemes needed to be developed. In particular, Ptolemy put forward a complicated system of 'wheels within wheels' which used very intricate geometric devices, called epicycles, for predicting the positions of each planet at any time. His system notably also reproduced retrograde motion.

Ptolemy's Model

Ptolemy went far beyond the scheme of the earlier Greeks, constructing a model out of circles and three geometrical devices, the eccentric and the epicycle, both of which had already been used by Hipparchus, and the equant, a device unique to Ptolemy. All of these devices were circular and therefore in keeping with the ancient tradition going back to Aristotle, Plato and even to Pythagoras.

One of the first challenges was to replicate the non-uniform motion of the Sun along the ecliptic on its annual 360-degree path across the background stars. If the path is divided into four 90-degree parts, the Sun would be at the zero point, on 21 March (when the Sun crosses the celestial equator in an ascending direction); 90-degrees further east on 21 June; 90-degrees farther still on 23 September and 90-degrees further on 22 December before then back in the starting point on 21 March, one full year later.

If the Sun moves uniformly, the times between these dates should be equal. But as can easily be determined by consulting a calendar, these times are not equal. The Sun takes a few days longer to move 90 degrees in spring or summer than it does in fall or winter. So any simple circular system based on motion with constant speed is flawed. As we have previously seen, Hipparchus, noting that the motion of the Sun and the Moon were not uniform, had already proposed devices to compensate for this apparent discrepancy, i.e. the geometric devices, the eccentric and the epicycle.

It is important to keep in mind that Plato, Aristotle and other ancient thinkers had decreed that a celestial object must move at a uniform rate and at a constant distance from the centre of the Earth. It is true, Ptolemy, like Plato, Aristotle and others, also believed that the Earth was at the centre of the universe, but unlike his forefathers, he did not insist that it stood at the geometric centre of all the perfect circles.

Instead, he proposed that the Earth could be off-centre in an eccentric position. Thus, motion that was really uniform about the centre would not appear to be uniform when observed from the Earth. By this logic the annual motion of the Sun, seen from the Sun, did not divide into four equal parts.

It is important to realize that Ptolemy was not just trying to explain the motion of the Sun and the Moon, but also the motion of all the other heavenly bodies, thus extending his work well beyond that of his predecessor, Hipparchus. The eccentric could be used to account for variations in the rate of motion of the planets as well as of the Sun and the Moon. However, though considered an important addition to the geocentric model, the eccentric alone could not describe changes as drastic as the retrograde motion of planets.

To account for this, Ptolemy made use of another device called the epicycle as shown in the figure below, which was also used by Hipparchus. Note that the planet is considered to be moving at a uniform rate on the small circle, the epicycle, whose centre moves at a uniform rate around the Earth on a large circle called the deferent.

It is a rather simple exercise to see that if a planet's speed around the epicycle is greater than its speed on the large circle

– the deferent – then the planet as seen from above would appear to move in a looping motion. When these loops are viewed from a location near the centre of the rotating system slightly edge-on, the motion looks like the retrograde motion of a planet in the plane of the solar system.

Ptolemy's system of epicycle on the deferent not only worked satisfactorily but also provided an unexpected bonus. The scheme explained not only retrograde motion but also the increased brightness of the planets when they are in the retrograde loop. A planet is on the inside of its epicycle during retrograde motion and thus would appear brighter to an observer from the Earth. This explanation no doubt confirmed that the model was a true explanation of what was taking place rather than just a geometric 'model'. It seems that both many of Ptolemy's contemporaries as well as generations to follow him believed he had surpassed the Greek goal to 'save the appearances' and had actually uncovered the true facts of the motion.

But there was another complication. Ptolemy could not synchronize the motions of the five planets exactly. For

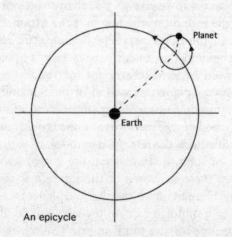

Epicycle travelling on a circular deferent.

example, the retrograde motion of Mars is not always the same angular size or the same duration. To allow for this variation, he brought in another geometrical device, the equant. This point, a variation of the eccentric, mirrors the Earth's position on the opposite side of the centre. Note the motion on the circle is not uniform around the centre. Instead (yielding to the ancient postulate), in order to incorporate the imperative to have some aspect of uniform motion in the system, Ptolemy proposed that motion is uniform as seen from this new point. The equant would eventually get Ptolemy into a great deal of trouble with future generations of astronomers who felt he had abandoned circularity.

Thus, Ptolemy took great liberties in adjusting the values of the epicycle, the eccentric, the deferent and the equant in developing his reproduction of the motion of the Sun, Moon and planets. Using seven different circles centred on seven different points near the Earth, Ptolemy succeeded in doing what his Greek ancestors had only dreamed of: to explain all the observed motion in the heavens in terms of circles and spheres. As the accuracy was considered astounding (for naked eye observers at least), his model was not challenged for many centuries.

Indeed it was not until the sixteenth century that anyone found serious discrepancies with Ptolemy's results. Up to that point in time, Ptolemy's values for the parameters of his geometric model – which predicted the positions of the heavenly bodies to the satisfaction of generations and generations of astrologers, navigators and astronomers – were tabulated, printed and made available. In order to use this complicated – albeit useful – scheme it was necessary to know the size of epicycle with respect to the deferent, the period of the motion of the planet in the epicycle and the period of the epicycle along the deferent.

The Almagest
Ptolemy's model was immortalized in his book, *The Almagest*, a thirteen-volume mathematical treatment of the phenomena

of astronomy that contains a myriad of information. Ranging from theories of the Earth to Sun, Moon and star movement as well as eclipses and a breakdown on the length of months, *The Almagest* also included a star catalogue containing forty-eight constellations, whose original names we still use today.

Though generally known as *The Almagest* (from a hybrid of Arabic and Greek, meaning 'the greatest'), Ptolemy originally titled his work *The Mathematical Syntax*, as he believed its subject, the motions of the heavenly bodies, could be explained in mathematical terms. In the work, Ptolemy synthesizes mathematical astronomy into a coherent whole, rendering his predecessors obsolete and dominating western and Islamic thought until the sixteenth century.

The opening chapters of *The Almagest* present empirical arguments for the basic cosmological framework within which Ptolemy worked. The Earth, he argued, is a stationary sphere at the centre of a vastly larger celestial sphere that revolves at a perfectly uniform rate around the Earth, carrying with it the stars, planets, Sun and Moon – thereby causing their daily risings and settings. Through the course of a year the Sun slowly traces out a great circle, known as the ecliptic, against the rotation of the celestial sphere. (The Moon and planets similarly travel backward and were therefore also known as 'wandering stars' rather than 'fixed stars'.)

In accordance with ancient Greek notions, the fundamental assumption of *The Almagest* was that the apparently irregular movements of the heavenly bodies were actually combinations of regular, uniform, circular motions. Yet how much of the published version of the tract is original is difficult to determine because almost all of the preceding technical astronomical literature is now lost. To his credit, Ptolemy attributed Hipparchus with essential elements of his solar theory, as well as parts of his lunar theory, while denying that Hipparchus constructed planetary models. He also made only a few vague and disparaging remarks regarding theoretical work over the intervening three centuries between Hipparchus and himself.

It may be surprising to many readers, however, that Ptolemy's reputation, especially as an observer, has been controversial since the time of the astronomer Tycho Brahe (1546–1601). Brahe pointed out that solar observations Ptolemy claimed to have made in AD 141 are definitely not genuine. There are, he continued, strong arguments for doubting that Ptolemy independently observed the more than 1,000 stars listed in his star catalogue. This charge of fraud has gained more support in recent years, especially after the important scholar, Robert R. Newton, concluded that Ptolemy faked his data to support his geometrical model. In 1977, Newton concluded in *The Crime of Claudius Ptolemy*: 'Ptolemy is not the greatest astronomer of antiquity, but he is something still more unusual. He is the most successful fraud in the history of science.'

Not all scholars agree with Newton, who, though a respectable historian of science, is considered something of a maverick by those who have examined his entire body of work. Though it is perhaps not particularly surprising that opposition to Newton exists, what is more remarkable is that the 1818 proof by French astronomer J.B.J. Delambre that Ptolemy lied about his observations of the equinox and the solstice was ignored for so long. Delambre crucially asked, 'did Ptolemy do *any* observing? Are not the observations that he claims to have made merely computations from his tables and examples to help in supporting his theories?'

Perhaps the final word on this issue should go to the dean of Ptolemaic scholars, Owen Gingerich. He states in his well-known article 'Was Ptolemy a Fraud?' that the Alexandrian's procedures in using observations available in AD 150 allowed him to correct some of the data. Gingerich points out that the correlation between the stated observations and the Ptolemaic theory could arise from deliberate observational selection and corrections to the data. Regrettably, Ptolemy did not bother to describe these 'corrections'. Convinced of the intrinsic correctness of this theory, Ptolemy probably replaced uncertain observations with what he perceived to be reliable measurements, that is to say, those that fit his model. In so doing he would not be the first theoretician who clearly expressed a belief in the primacy of theory over experiment or in believing that theory represented nature better than individual observations, which can easily be affected by human error.

Whatever the truth may be, it is nevertheless clear that Ptolemy was pre-eminently responsible for the geocentric cosmology that prevailed in medieval Europe. This was not due to *The Almagest*, as there was no description of the model in that book, but rather in a later treatise now known as *Planetary Hypotheses*. Scholars have confirmed Ptolemy's authorship of this latter book in which his system, involving at least 80 epicycles, explains the motions of the Sun, the Moon and the five planets that were known during his time. He believed that the planets and Sun orbit the Earth in the following order: Mercury, Venus, Sun, Mars, Jupiter, Saturn. This became known as the Ptolemaic system.

In all of the criticism of Ptolemy, what is not disputed is the mastery of mathematical analysis that he exhibited. After completing his geometric model, he computed the numerical tables, which are published in *The Almagest*, which enable planetary positions and other celestial phenomena to be calculated for arbitrary dates. This had a profound influence on medieval astronomy, in part through a separate, revised version of the tables that Ptolemy published as *Procheiroi kanones* ('handy tables'). He later taught astronomers how to use dated, quantitative observations to revise cosmological models.

Gingerich, who perhaps knows Ptolemy best, dismisses the moral question of Ptolemy's handling of ancient data and challenges historians to reconstruct his 'pioneering trail to the most complete mathematical achievement in ancient astronomy'. He considers Ptolemy 'the greatest astronomer of antiquity'.

4

CROSSING THE DARK AGES

For nearly forty centuries, man has been investigating the heavens – from the intellectual pursuits of Mesopotamia in about 2000 BC to the Cosmic Background Explorer satellite that was launched in 1989. But this process has not been one of continuous steady progression. Implanted within this vast period of human intellectual history, there exists a span of approximately fourteen centuries, just about one-third of the total time period, when nothing happened to further the understanding of the movement of bodies in the heavens. In other words, astronomy fell into serious decline.

With his masterwork *Mathematike Syntaxis* – or to use the Latin version of its Arabic name, *The Almagest* – Ptolemy crystallized the Greeks' compulsion to 'save the appearances'. In keeping with their convention, he established that predicting the position of the planets was far more important than understanding any other aspect about the planets, stars, Sun and Moon. In their critique of Ptolemy's complicated mechanisms (for example the 'wheels within wheels' that we have seen previously), Ptolemy's contemporaries, along with the generations that followed, worried more about philosophical principles such as uniform circular motion rather than any real physical characteristics such as the composition of the planets or their true motion in the heavens.

The sensational, agonizing rejection of this approach, initiated in the mid-fifteenth century in what is now called the scientific revolution, stands out as that part of the Renaissance which was not a 'rebirth' of the ideas of the ancients, but rather the opposite. Beginning with Nicolaus Copernicus in 1543, then Tycho Brahe, Johannes Kepler and finally Galileo Galilei, the goal of this movement was to eliminate the incorrect postulates of Aristotle from existing models of the heavens. This was not an easy task, given the Greek philosopher's importance in the fields of metaphysics, rhetoric, poetics, logic and science. Though he laid the foundation of subsequent studies in, for example, biology, Aristotle's physics and cosmology were ultimately found to be faulty.

There has, however, always been a mysterious chasm separating the Greeks and the heroes of the scientific revolution – Copernicus, Tycho, Kepler and Galileo. In order for the revolution to take place in the sixteenth century, at least a copy of Ptolemy's *Almagest* (amongst other ancient works surviving the so-called Dark Ages) had to have landed in the hands of men like the Polish cleric Nicolaus Copernicus. Copernicus would eventually make a massive assault on the ancient theories by placing the Sun at the centre of the known universe. Tracing this leap of fourteen centuries may seem a complicated and fascinating task, but in reality, it is quite simple. The story begins with Mohammed and the rise of Islam.

With the death of the great prophet in AD 632, the Muslim world began a period of rapid expansion. Under the rule of the first caliphs, Muslim armies began assaulting the borders of both Persia and the Byzantine Empire. In the latter case, they set their sights south towards the rich provinces of Byzantine Africa, especially Egypt, which had been part of the former Roman Empire. The Muslim invasion of Egypt was easily accomplished because the caliph's forces were larger than any army the Byzantines could field at the time. The destruction of Byzantine military power at the ensuing battle of Heliopolis in the summer of AD 640 and the victory over the Byzantine defenders at Babylon effectively broke the Byzantine power hold in Egypt. The city of Alexandria was left defenceless, with only a fraction of provincial forces garrisoned in the city itself. On 8 November 641, after a fourteen-month siege, Byzantine officials at last capitulated, turning the city over to Muslim hands.

Culturally, the city continued to function in much the way that it had under Byzantine rule and Greek, Coptic and Arabic were all spoken fluently throughout the city. With a tolerant Muslim administration, documents continued to be published in Greek and Coptic for some time following the takeover. These included documents about medicine, mathematics and alchemy, whose practices thrived under the budding advances of Islamic intellectualism.

Later in the eleventh century, Arabic replaced Greek and Coptic as the principle language of Alexandria. Though these Arabic conquerors are historically blamed for the final destruction of the great Library, finishing what the Romans started with Caesar in 48 BC and the Emperor Aurelian in the third century, it is obvious that many of the great works in the collection were at the time translated into Arabic. Although the actual circumstances and timing of the physical destruction of the Alexandrian Library remains uncertain, it is clear that by the eighth century, this historic institution was no longer significant. It had ceased to function in any important capacity and the city was no longer a major research centre for the Islamic world.

The Muslim population continued to migrate, and with them travelled their Arabic translations of many of the great works of the Ancient World. By AD 711, the Arabs and Moors had arrived in Andalucía, the southern part of the Iberian Peninsula, where they rapidly established control. They were to remain until 1492 when they were famously expelled by the Catholic monarchs, Ferdinand and Isabella. Yet, as early as the tenth century, Christians had launched the Reconquista, an attempt to regain control of Spain from the Muslims.

In certain major Spanish centres of Islamic culture, like Cordoba and Toledo, Christian traditions continued to survive and Islamic intellectual pursuits were appreciated at the same time. Thus, the Christian Reconquista was more of an assimilation, a fusion of different races and religions. Whenever the Christian rulers needed a surgeon, an architect, or even a dressmaker, they applied to Cordoba and were as likely to be presented with a Moor or a Jew as with a Christian. So over a period of about three hundred years, the 'barbaric West' was given glimpses of the progress that was being made in Spain. Despite these glimpses, the full extent of Islamic achievement was not realized until Toledo was retaken in 1085 and King Alfonso VI discovered its buildings, libraries and universities intact.

Toledo soon became a centre of multilingual culture, with a large population of Arabic-speaking Christians, known as Mozarabs, as well as Jews, Arabs and Greeks. Toledo's tradition of scholarship, and the books that embodied it, survived the Christian conquest of the city in 1085. A further factor in the preservation of this tradition was that Toledo's early bishops and clergy came from France, where Arabic was not widely known. The cathedral consequently became a centre for translations, which occurred on a scale of importance that has no match in the history of western culture.

This progress in Toledo is at odds with the common idea of the Dark Ages, the period between the collapse of the Roman Empire and the re-creation of political stability and written historical records in Europe. Typically characterized by a lack of central control as well as ignorance, this period is often dismissed as an intellectual black hole. However, thanks to the Arab migration, Spain was the most strategically placed country to weather the three hundred years that were the Dark Ages of Europe. In current historical nomenclature, this period is called the Early Mediaeval or Early Middle Ages.

One of the sponsors of translations in Spain was Archbishop Raymund of Toledo (1125–52). A Benedictine monk, born in French Gascony, he promoted the translation of scholarly texts from Arabic into Latin during his episcopacy. Nineteenth-century historians have even gone so far as to propose that Raymund established a formal translation school. However, no specific evidence for such a school has emerged, and its existence is now doubted.

Nonetheless, Raymund recruited scholars from throughout the Christian world. Among them was a young Italian, Gerard of Cremona. Gerard was the most productive of the Toledo scholars and translated 87 books, mostly from the ancient world. These included Ptolemy's *Almagest*, Aristotle's *Physics* and *On the Heavens*; Euclid's *Elements of Geometry* and Archimedes' *On the Measurement of the Circle*.

Born in Cremona, Gerard was dissatisfied with the meagre philosophies of his Italian teachers, and so decided to travel to Toledo. Though no detailed information exists of the date when Gerard went to Castile, it was no later than 1144. In Toledo, he learned Arabic, initially so that he could read Ptolemy's *Almagest*, which retained its good reputation among scholars despite the lack of a Latin translation. Gerard's story is typical of many of the scholarly pilgrims to Toledo:

'. . . [He] arrived at a knowledge of each part of [philosophy] according to the study of the Latins, nevertheless, because of his love for the Almagest, which he did not find at all amongst the Latins, he made his way to Toledo, where seeing an abundance of books in Arabic on every subject, and pitying the poverty he had experienced among the Latins concerning these subjects, out of his desire to translate he thoroughly learnt the Arabic language. . .'

Gerard of Cremona's Latin translation of the Arabic text of Ptolemy's *Almagest* remained the only version that was known in western Europe for centuries until Johannes Regiomontanus, a contemporary of Copernicus, translated it from the Greek original in the fifteenth century. *The Almagest* formed the basis of a mathematical astronomy until it was eventually eclipsed by the heliocentric theory of Copernicus.

In tracing the transmission and translation of ancient texts from Alexandria to Toledo and the rest of Europe, the mystery of how Copernicus was able to obtain a copy of Ptolemy's *Almagest* has been solved. Copernicus studied as a young man in the northern cities of Bologna and Padua, not far from the hometown of the prolific Gerard, the early translator of Ptolemy's great work.

Thomas Aquinas

No discussion of the assimilation of the works of Ancient Greece into Europe is complete, however, without a description of the role of Thomas Aquinas in promoting Aristotle's writings within the Catholic Church.

Thomas Aquinas was an Italian thirteenth-century priest of the Roman Catholic Church in the Dominican Order. An immensely influential philosopher and theologian in the tradition of scholasticism, he was the foremost classical proponent of natural theology. His influence on western thought is considerable and much of modern philosophy was conceived as a reaction against, or as an agreement with, his ideas, particularly in the areas of ethics, natural law and political theory. Aquinas viewed theology, or sacred doctrine, as a science – the raw material data which consists of written scripture and the tradition of the Catholic Church. Faith and reason, while distinct but related, are the two primary tools for processing the data of theology. He believed both were necessary – or, rather, that the confluence of both was necessary – for one to obtain true knowledge of God. According to Aquinas, God reveals himself through nature, so to study nature is to study God.

It can be said with certainty that the greatest influence upon the thought of Aquinas was the philosophy of Aristotle whom he simply referred to as 'The Philosopher'. How Aquinas came to know Aristotle's works is important in the intellectual history of the West.

As we now know, after the fall of Rome Justinian closed Plato's Academy and the Lyceum of Aristotle in AD 529.

The majority of the most important texts of Greek philosophy therefore became unavailable. But Islamic scholars in the near East saved many of these ancient manuscripts they had found in Byzantine libraries and from the richest library in the ancient world, the library at Alexandria.

By the twelfth century, these manuscripts (along with the commentaries made on them) made their way back into Europe by way of Andalucía, Sicily and North Africa. This was due to the reactivation of trade routes, which the end of the Crusades made possible. These texts helped to make the twelfth century Renaissance a reality and by the middle of the thirteenth century, French and Italian universities became inundated with these ancient texts, especially the philosophical works of Aristotle.

Because Aristotle's method for choosing fundamental principles was not based on observations of any kind, science in the Middle Ages became a theoretical subject and a branch of philosophy that was sometimes referred to as 'natural philosophy'. Although the twelfth-century philosopher Roger Bacon and fourteenth-century philosopher Nicole Oresme praised the concept of experience and experimentation long before Tycho Brahe measured the heavens in the sixteenth century, these were not the typical activities of a natural philosopher.

Aristotle's methodology included applying a few fundamental principles to the motion of simple objects. However, the manner in which he selected these principles was not in any way scientific because he assumed that they could be intuitively perceived as self-evident truths, not through experiment or observation. For example, material objects came to rest in relation to the cosmic centre, which was the Earth. Thus an object fell because of its desire to be at the cosmic centre. A heavy object, with its greater density, would therefore fall faster than a light object. After performing simple experiments, Galileo would find this to be completely false. Also, Aristotle's scheme proposed that in the perfect heavens, objects would move at a uniform speed in that most perfect of figures, the circle.

Greek science's fatal flaw dominated the study of nature until the Renaissance. Mostly, this was because there was no mechanism to produce consensus, the lifeblood of the success of modern science. Thanks to Aristotle and others, Greeks saw tests of scientific conclusions as no more necessary than tests of politics or aesthetics. Conflicting views could be argued indefinitely and the idea of something like a 'critical experiment' would never have occurred.

Later, Aquinas fitted Aristotle's complete *oeuvre*, including his physics and cosmology, together with the Church's moral and spiritual doctrine to create a compelling synthesis. In this way, concepts like motion toward the centre of the Earth, perfect uniform circles in the heavens and geocentricism became part of the Church's teaching throughout the Christian world. This was the context in which Ptolemy produced his masterwork of 'circles within circles' – with no one ever systematically checking parameters or amending models to improve accuracy. This lasted for centuries. From the death of Ptolemy until the birth of Copernicus in 1473 there exists in the West not more than a dozen new records of accurate planetary positions.

In the eastern part of the Muslim world astronomers did begin to experiment with new planetary models by adding one or two more epicycles. But unlike the modern viewpoint, these innovations were not introduced to improve accuracy but rather to satisfy further philosophical requirements. These little 'epicyclets' were therefore designed to replace the equant, the off-centre epicycle, in order to preserve the ancient requirement of uniform circular motion and to set up a purely mechanical model of the planetary system.

The powerful influence of 'The Philosopher' during the ancient and medieval worlds coupled with the persuasive skill of his disciple Aquinas, resulted in the Church adopting Aristotelianism as an integral part of its teaching. Consequently, Aristotelians at the universities were protected – and it was not until the Renaissance that they began to come under threat.

PART II
NEWTON'S UNIVERSE

5

NICHOLAUS COPERNICUS: INNOVATOR OR TRADITIONALIST?

It is fashionable today in certain circles of historical writing on physical science to play down the importance of the scientific revolution that occurred during the sixteenth and seventeenth centuries. Yet the remarkable contributions of five men, each from a different country in Europe – namely Poland, Denmark, Germany, Italy and England – changed our understanding of the universe unequivocally and dramatically in a period of only 144 years. So it seems sensible to use the designation 'revolution' to describe the effects of works such as Copernicus' *De Revolutionibus* (1543) and Newton's *Principia* (1687), since ideas on motion in the heavens and on the Earth were completely transformed. Remarkably, these ideas remain intact in the twenty-first century, except in situations of very high velocity and extreme gravity which were modified by Einstein's two theories of relativity in the early twentieth century.

As we have seen, the important works of the ancient world, including the writings of Aristotle, were transmitted to European philosophers via Arabic translations of the originals. These documents surfaced in the rich Muslim culture of Andalucía in the twelfth century. After being translated again from Arabic back into Latin, these books made their way from Spain to Italy and became available to the next hero of the 'discovery of the universe', Nicholaus Copernicus.

Only certain sporadic facts about Copernicus' early life are known; sadly, a biography written by his ardent disciple Georg Joachim Rheticus remains lost. However, according to a contemporary horoscope, it is known that Copernicus was born on 19 February 1473, in Torun, a city in north-central Poland on the Vistula River, south of the major Baltic seaport of Gdansk. His father was a well-to-do merchant and his mother, Barbara Watzenrode, also came from a leading merchant family. After his father's death, sometime between 1483 and 1485, Nicolaus, the youngest of four children, was taken under the protection of his mother's brother Lucas Watzenrode.

Between 1491 and 1494 Copernicus studied liberal arts at the Jagellionian University in the Collegium Maius, a facility that had excellent specialists in astronomy and mathematics. However, he left before completing his degree and travelled to Italy where he resumed his studies at the University of Bologna, one of the oldest and most famous universities in Europe. Founded in the eleventh century, Bologna became the principal centre for studies in civil and canon law in the twelfth and thirteenth centuries. The university attracted students from all over Europe

Copernicus' uncle had obtained a doctorate in canon law at Bologna in 1473, and as a young student, Copernicus hoped to pursue the same course of studies. Nicholaus' time at Bologna (1496–1500) was short but significant. By chance, he lived in the same house as the principal astronomer at the university, Domenico Maria de Novara. Born in Ferrara in 1454 and a professor for twenty-one years at Bologna, Novara had the responsibility of issuing annual astrological prognostications for the city, perhaps for economical reasons as was common in those times. Copernicus soon became Novara's assistant and helped with the production of the annual forecasts. In 1497, only a few years after Christopher Columbus arrived in America, Copernicus was checking the new and full-Moon times derived from the commonly used Alfonsine Tables. These were then used in Novara's forecast for the year 1498.

It is also recorded that Novara introduced the young law student to two important books that were to frame his future as a student of the heavens. The first, *Epitome of Ptolemy's Almagest* was written by Johann Müller, a German writer known as Regiomontanus, a man whom Novara declared with pride as his teacher. The second was *Disputations Against Divinatory Astrology* by Giovanni Pico della Mirandola. The first provided a summary of the foundations of Ptolemy's astronomy and contained critical expansions of certain important planetary models which may have helped inform Copernicus' heliocentric hypothesis.

Pico's *Disputations* offered a devastating sceptical attack on the foundations of astrology. Among his criticisms was the charge that astrologers could not be certain about the strengths of the powers issuing from the planets because astronomers disagreed about the order of the planets. This book influenced the young student at this critical juncture in his education. Pico's book also had an influence on other astronomers such as Tycho Brahe and Johannes Kepler, and caused them to question the 'scientific' aspects of astrology. One can assume the effect on Copernicus was similar.

Copernicus quickly began to disregard canon law in favour of more scientific pursuits. In 1500, in the midst of the High Renaissance in Italy, he spoke before an interested audience in Rome on mathematical subjects, and between 1501 and 1503 studied medicine at the University of Padua. At this time medicine was closely allied with astrology as the stars were thought to influence the body's dispositions. Thus, Copernicus' astrological experience at Bologna was better training for medicine than one might imagine today. It is about this time that he probably came in contact with Ptolemy's *Almagest,* as it was readily available in European universities in Latin translations from Arabic.

In May 1503, Copernicus received a doctorate – like his uncle Watzenrode – in canon law. He received the degree from an Italian university where he had not studied but transferred into at a late stage: the University of Ferrara. By the time he returned to Poland, his uncle had become a Bishop and was able to arrange a lifetime position for him at the cathedral of Frombork. Copernicus' actual duties at the bishop's palace were largely administrative and medical. As a church canon, he collected rents from church-owned lands, secured military defences, oversaw chapter finances, managed the bakery, brewery and mills and cared for the medical needs of the other canons and his uncle.

This background seems hardly compatible with a career as an astronomer. In fact, Copernicus's astronomical work took place in his spare time. He considered the study of the heavens to be a hobby and dabbled for forty years developing his own Sun-centred geometric system of the world. As a Catholic, he knew the Church would have denounced as heretical any attempt to remove the Earth from the centre of the world, so he was in no hurry to publish. In addition, he was actually in a Protestant, Lutheran, country though protected by the Alps from the excesses of the Roman authorities.

At Frombork, Copernicus received an adequate income, which was derived from peasants working the farmlands under his administration, and a lifetime tenured position. He was therefore able to afford to indulge his astronomical interest in the structure of the universe. He probably hit upon his main idea of heliocentrism sometime between 1508 and 1514 and during those years wrote a manuscript usually referred to as the *Commentariolus* (Little Commentary). A short, hand-written document distributed anonymously to friends with astronomical interests, the *Commentariolus* is very important in understanding the canon's interests.

Here Copernicus postulated that if the Sun is assumed to be at rest and the Earth is assumed to be in motion, then the remaining planets fall into an orderly relationship. Their sidereal periods (the time it takes for them to circle the Sun as measured from the Earth) increase in magnitude as they recede from the Sun. Consider the following example: Mercury takes 88 days to circle the Sun; Venus 225 days; the Earth 1 year; Mars 1.9 years; Jupiter, 12 years and Saturn 30 years. Years later, Copernicus would calculate these periods using observations dating back to antiquity.

Copernicus was aware that his contemporaries would not accept his groundbreaking idea with ease. Consider the following quote from his late work, *De Revolutionibus* (1543):

> The ideas here stated are difficult, even almost impossible, to accept; they are quite contrary to popular notions. Yet with the help of God, we will make everything as clear as day in what follows, and these for those who are not ignorant of mathematics . . . the first and highest of all the spheres is the sphere of the fixed stars. It encloses all the other spheres and is itself self-contained; it is immobile; it is certainly the portion of the universe with reference to which the movement and positions of all the other heavenly bodies must be considered. If some people are yet of the opinion that the sphere moves, we are of contrary mind; and after deducing the motion of the earth, we shall show why we so conclude. Saturn, first of the planets that accomplishes its revolution in 30 years is nearest to the first sphere. Jupiter, making his revolution in 12 years is next. Then comes Mars revolving once in two years. The sphere which contains the Earth and the Moon and which performs an annual revolution occupies the fourth place in the series. The fifth place is that of Venus, revolving in nine months. Finally, the six places occupied by Mercury revolving in just 80 days. In the midst of all, the Sun reposes, unmoving.

At Frombork, Copernicus again tried to justify his new way of thinking:

> For a long time I reflected all the confusion in the astronomical traditions concerning the derivation of the motion of the spheres

of the universe. I began to be annoyed that the philosophers have discovered no sure scheme for the movements of the machinery of the world, created for our sake of the best and most systematic Artist of all. Therefore, I began to consider the mobility of the earth and even though the idea seemed absurd, nevertheless I knew that others before me had been granted the freedom to imagine any circles whatsoever for explaining the heavenly phenomena.

In addition to his own research, his studies uncovered other writers who had considered the movement of the Earth, like the second century Athenian philosopher Plutarch and the Roman orator Cicero. He thus felt validated in the concepts he was developing towards his own theory. Since his work was not in the public domain, and he held a position as a canon in a small cathedral precinct, Copernicus did not promote himself as an astronomer. Still, his reputation outside local Polish circles as a stargazer of considerable ability is evident from the fact that in 1514 he was invited to offer his opinion on the critical problem of reform of the calendar at the Church's Fifth Lateran Council.

We can therefore be fairly certain that Copernicus explored the accepted system of the heavens – that of Ptolemy – during his studies (of canon law) in Italy at the peak of the High Renaissance and found it wanting. He then spent a long career as a churchman in a provincial parish in Poland whilst secretly working on the orbits of planets, the movement of the Sun and the Moon and other questions before finally publishing his work in what was to become one of the most famous books ever printed. But what exactly motivated Copernicus to spend forty years developing an alternative to Ptolemy's Earth-centred model – a model that seemed satisfactory to most astronomers and astrologers in the early part of the sixteenth century?

Was the main purpose of his introducing a new system to improve the accuracy of the predictions of the planetary positions? Or was he attempting to get closer to the Platonic ideal by using new combinations of circles resulting in fewer uniform circular motions? Or did he wish to eliminate the troublesome equants introduced by Ptolemy? Or lastly did he propose his new system on aesthetic grounds of greater simplicity and beauty?

If his own writing is to be believed, Copernicus was indeed concerned with Plato's old problem, that is, to construct a system by combining the fewest possible uniform circular motions. Furthermore, it does appear that he was trying to rid the Ptolemaic system of that which seemed so contrary to Plato's assumptions, that is the unpopular equant. As early as 1512 he wrote in the *Commentariolus*:

> ... the planetary series of Ptolemy and most other astronomers, although consistent with the numerical data, seem likewise to present certain difficulties. For these theories were not adequate and unless certain equants were also conceived; it then appeared that a planet moves with uniform velocity neither on its deferent nor about the centre of its epicycle. Hence a system of this sort seemed neither sufficiently absolute nor sufficiently pleasing to the mind. Having become aware of these defects, I often consider whether there could perhaps be found more reasonable arrangement of circles, from which every apparent inequality would be derived and in which everything would move uniformly out of its proper centre.

Copernicus is often described as a revolutionary figure in the history of astronomy, but in many ways he is the last representative of the old Ptolemaic tradition rather than a harbinger of the new. He laboured for forty years on his model, adding and removing circles (much as Ptolemy did) in an attempt to explain the structure of the universe and the motion of the planets.

Copernicus' new model was entirely within the Ptolemaic tradition – he completed his system by appropriating over thirty epicycles and eccentrics. His approach included many of the same philosophical methods of the Alexandrian of fourteen centuries earlier, and in many ways appeared to be writing a new version of *The Almagest*. Clearly, he was compelled by his background to consider only the constructions consistent with the philosophy of the ancients, particularly Plato and Aristotle.

However, despite the fact that Copernicus added no new physics and seemed intent on making merely another calculating machine to predict the position of the heavenly bodies at any arbitrary time, he did propose a whole new world view by placing the Sun at the centre of his system. In his model, all of the planets – including the Earth – revolved around the Sun.

Maybe it is not fair to minimize Copernicus' contribution to the eventual development of the Newtonian solution to the problem of motion in the solar system. Though he was caught up in the use of the same non-physical devices – epicycles – invented by Ptolemy, his approach can seem quite modern in twenty-first century terms if we consider the assumptions on which his model was based. True, some of these merely mimic Aristarchus, but the Polish clergyman was very clear and precise in his postulates and that, in a sense, is very modern indeed. He wrote:

- There is no one precise, geometrical centre of all the celestial circles or spheres.
- The centre of the Earth is not the centre of the universe, but only of gravitation and other lunar sphere.
- All the spheres revolve around the Sun and therefore the Sun has a central location in the universe.
- The distance from the Earth to the Sun is very small in comparison with its distance to the stars [this explains why there is no parallax].
- Whatever motion appears in the sky arises not from its motion but from the Earth's motion.
- The Earth together with its water and air performs a complete rotation on its fixed poles in a daily motion, while the sky remains unchanged.

- What appears to us as the motion of the Sun arises not from its motion but from the motion of the Earth and we revolve about the Sun like any other planet; the Earth has then, more than one motion.
- The apparent retrograde motion of the planets arises not from their motion but from the Earth's. Motions of the Earth alone therefore are enough to explain so many apparent motions in the sky.

In addition, there are two major simplifications of the Copernican system over the Ptolemaic scheme, which must have convinced many contemporaries that Copernicus had made not just a better calculator for predicting planetary positions, but had indeed discovered the true structure of the world. The first was an explanation for the orbits of Mercury and Venus, the morning and evening stars, which are bright objects that sometimes rise just before or after the Sun. That is because these are always close to the Sun.

For Copernicans, this phenomenon was easy to understand. Both of these were 'interior' planets and circled close to the Sun. When viewed from the Earth, they follow close behind the rising or setting of the Sun. The Ptolemaic system considers that if Mercury, Venus and the Sun travel independently with separate orbits, they should appear to be widely separated at some point in time. Ptolemy thus must have assumed these planets were somehow attached to the Sun, travelling with it. Why these two planets were singled out to be tied to the Sun while the others were free to roam remained unanswered by Ptolemy.

The second simplification is even more dramatic. Copernicus was able to explain the retrograde motion of some planets without the use of an epicycle as shown earlier in the figure in the section on Artistarchus. It is evident that as the faster moving Earth passed one of the slower moving 'exterior' planets, for example Mars, the slower planet would appear from the Earth to back up and then move forward again as the Earth passed it in its orbital motion. These demonstrations must have been reassuring for the Copernicans to point to when discussing the relative merits of the two systems.

We know today that these differences between Copernicus and Ptolemy depend on the frame of reference. Ptolemy was standing on the Earth when making his observations and assumed that the Earth was motionless. Copernicus preferred to believe that he was somehow looking from the Sun. His point of view was based on watching the planets swirl in orbits 360 degrees around the Sun, that fiery star which illuminates and powers planets with its radiation.

Today, in spite of our understanding of the arrangement of the heliocentric solar system, all stargazers, whether amateur or professional, still use the Ptolemaic viewpoint (a perspective taken from a stationary Earth). This is not surprising as coordinates and subsequent measurements are much less complicated when taken from the geocentric point of view. However, computing the position of the planets as seen from the Earth is still a challenging prospect.

It was not surprising that Copernicus was reluctant to publish his book. In the mid-sixteenth century, the notion of the moving Earth seemed somewhat absurd. First of all, medieval cosmology and physics was based on the idea that the Earth was the centre of the universe. This assumption was a starting point for the great philosophers, including Aristotle, who postulated that objects fall to the Earth because, as we have seen before in his work *On the Heavens*, 'objects tend to fall to the centre of the universe'. Additionally, rapid motion of the Earth (now known to be about 18 miles (29 kilometres) per second in the orbit around the Sun) would elicit severe criticism as it was thought at that time that objects would fly off the surface of the Earth.

De Revolutionibus

To some, the life of Copernicus may seem quite dull: a church official secretly writing his astronomical masterpiece over almost four decades in a corner of a tower next to his parish church. But a great deal of excitement is associated with the publication of his great work, which did not occur until he was near death. Strangely, the most interesting part of his life story could well be the intrigue surrounding the actual publication of his *magnum opus* in 1543, for its publication is shrouded in exceptional deception and betrayal amongst his colleagues.

The writing was hidden for some thirty-six years. Copernicus himself was only given a printed copy of his manuscript a few hours before he died. But before it was printed, a few bizarre characters managed to find out about his work and affect its publication. The first, a twenty-five-year-old Lutheran admirer, was an academic from the University of Wittenberg named Georg Rheticus. He came to stay with Copernicus in 1539 and stayed almost three years, between 1539 and 1542. During that time, he managed to convince the older man into publishing the main elements of the heliocentric hypothesis as part of a minor document called *Narratio Prima,* in 1541. Rheticus even agreed to put his name on the piece instead of Copernicus though it was a joint endeavour.

Narratio, a summary of the theoretical principles of the model, was a kind of 'trial balloon' for what was to come in the full treatise. Frustrated that the work was still mostly unknown, Rheticus elicited Copernicus' consent to full disclosure by using a clever pretext. Rheticus reasoned that the full work would demonstrate the value of the new theory for computing more accurate planetary tables. In addition, Copernicus' work could be presented only as modification of Ptolemy's.

Rheticus was anxious to get the work into print and when the manuscript was completed, he ignored the reluctance of the author, who by now must have been frail and retiring. When Rheticus left Frauenberg to return to his teaching duties at Wittenberg, he took the precious text with him in order to arrange the publication at Nurenberg at one of the best printers in Germany. However, finding himself to be too busy to arrange the printing himself, he turned the task over to Andreas Osiander, a theologian with valuable experience of getting books through production.

But in an outrageous breach of confidence and propriety, Osiander added an unsigned 'letter to the reader' right after the title page. This note ultimately undermined Copernicus' belief that his scheme was true and correct, and suggested that the model was merely hypothetical. Without referring to the author or his acolyte Rheticus, Osiander wrote that there was no pretence of truth in the book and that astronomy was, in fact, incapable of finding the causes of heavenly phenomena. The unsuspecting Rheticus and Copernicus found themselves double-crossed. Moreover, since the authorship of the letter was not indicated in the printing, the letter became the preface to *De Revolutionibus*. Osiander was probably motivated to appease church authorities in hopes that they would not block the publication of what was then a heretical viewpoint.

The identity of the rogue writer was not known until over sixty years later when Kepler revealed what was generally the best-known secret among sixteenth-century astronomers. Though Osiander's actions were surely unacceptable with regards to preserving the original characteristics of the work of an author, it did make it possible for *De Revolutionibus* to be read as a new method of calculation rather than as a work of natural philosophy. In so doing, this may have contributed to the initial positive reception of the book. Perhaps this was Osiander's intention after all.

Legend has it that a copy of *De Revolutionibus* was placed in Copernicus' hands just after he lost consciousness from a stroke in Frombork on 24 May 1543. Even more romantic is the postscript to this tale, for it is said that Copernicus later awoke long enough to realize that he was holding in his hands his great work and . . . then he expired. This may be apocryphal but nevertheless will forever take its place as part of the folk-lore of the history of astronomy.

As Copernicus expected, his theory was denounced as 'false and altogether opposed to the Holy Scriptures'. For had it not been written in the Bible that the Sun moved:

On the day the Lord gave the Amorites over to Israel, Joshua said to the LORD in the presence of Israel: 'O sun, stand still over Gibeon, O moon, over the Valley of Aijalon.' So the sun stood still, and the moon stopped, till the nation avenged itself on its enemies, as it is written in the *Book of Jashar*. The sun stopped in the middle of the sky and delayed going down about a full day. . .

There is an unusual coda to the publication of this historic book. According to another provocative and unsubstantiated comment made by the science historian Arthur Koestler in his well-known volume, *The Sleepwalkers*, *De Revolutionibus* was not widely read at the time of its first publication. However, Owen Gingerich, a widely recognized authority on Copernicus, was sceptical of this remark and undertook a thirty-five year-long project to examine every surviving copy of the first two editions of the work. After a heroic effort of dogged scholarship, in most cases personally examining the rare copies, he disproved Koestler's assumption, showing that annotations in the margins of these now expensive collector items proved that contemporaries had indeed read *De Revolutionibus*. Starting with a heavily annotated copy he found in Edinburgh, Gingerich was encouraged at the evidence of an expert readership and thus concluded that there was great interest and debate about the book.

This research earned Gingerich the Polish government's Order of Merit in 1981 and his efforts and conclusions were published in a book for general readers entitled, *The Book Nobody Read* (2004). Ironically, due largely to Gingerich's work, *De Revolutionibus* is now researched and catalogued better than any first-edition historical text in existence except the original Gutenberg Bible. As regards the value of a first edition copy of this famous work, a recent sale at Christie's of important scientific books brought a sale price in excess of US$2 million.

6

RENAISSANCE ASTRONOMY: TYCHO BRAHE AND JOHANNES KEPLER

In August 1563 a young Danish student, aspiring to be an astronomer, first noticed a significant discrepancy between his own observations of certain planets and the predictions of the contemporary planetary tables of the time, the Alfonsine and the Copernican ephemerides. His name was Tycho Brahe. Here is a report in his own words, taken from a short autobiography that appears in his well-known later work, *Mecanica* (1598):

In year of our Lord 1563, on the occasion of the great conjunction of the upper planets, which took place at the end of Cancer and the beginning of Leo, I had reached the age of sixteen years and was occupied with studies of classical literature in Leipzig where I lived with my governor. During the evenings I began to study astronomy more and more with the aid of a few books, particularly Ephemerides which I bought secretly in order that the governor should not become aware. Having a natural inclination for the subject, I quickly got accustomed to distinguishing the constellations of the sky and in the course of a month I got to know them all.

Soon my attention was drawn towards the motion of the planets, but when I noted their positions on the fixed stars using lines drawn between them, I noticed that their positions agreed neither with the Alfonsine tables nor the Copernican tables, although the agreement of the latter was better. I began to make measurements with a crude instrument I constructed myself, somewhat like a compass, and though this was not very accurate it became quite clear to me that both tables suffer from intolerable errors. This was actually apparent from the great conjunction of Saturn and Jupiter in the year 1563, which I mentioned at the beginning and this discrepancy is precisely the reason why I became interested. The discrepancy was a whole month for the Alfonsine numbers even several days compared with those of Copernicus.

From that moment 1563, through the whole development of the scientific revolution and into the modern era, *observation* has guided *speculation* in the evolving picture of the universe. From Tycho's self-made crude instrument in Leipzig to Galileo's refinement of the telescope; from the advent of the spectroscope to the orbiting of the Hubble Space Camera; from simple radiometers to the x-ray detectors studying black holes, new phenomena have been followed by new theories of explanation that continually transform our understanding of the universe. It is not an overstatement to suggest that the primacy of observations over theory that has characterized science for four hundred years began with Tycho in 1563.

Tycho is famous to many thousands of casual followers of the history of astronomy as the eccentric aristocratic who lived on a desolate island and famously wore a metal plate on the bridge of his nose. Only a few of those who have heard of the eccentric Dane know anything at all about his contribution to the history of the discovery of the universe. In fact, the significance of the contributions of Tycho to the story of the discovery of the universe cannot be over emphasized.

Nevertheless, to all but historians and astronomers, his importance is generally unknown. He didn't invent a new solar system like Copernicus or a new force, like gravity, as discovered by Newton. But in concert with his protégé Kepler, he started the new physics of the sky based on observation and instrumentation rather than theory.

The meeting of these two eccentric characters at the very beginning of the seventeenth century is remarkable in its providence. The symbiosis was perfect as one could not excel without the other. And it began that evening in 1563, when Tycho first noticed the large discrepancies between his observations and the predictions recorded in Ptolemy's (from AD 150) and Alfonso X's (from AD 1250) tables.

Life of a Noble Astronomer

Tycho was born three years after Copernicus received the first copy of *De Revolutionibus* on his deathbed. A self-determined and confident stargazer from his university days, in 1563 he found himself frustrated by the inaccurate prediction of a Jupiter-Saturn conjunction which was the result of using the accepted tables of Ptolemy, now over one thousand years old. The young nobleman knew he could do better and secretly began recording the positions of the planets in the night sky when he was supposed to be studying jurisprudence. He pursued his interest despite strong resistance from his aristocratic family, who expected him to follow a career as a court diplomat – not a lonely stargazer.

After his foster father died, however, he became fiercely independent, travelling in Germany and Switzerland and meeting other astronomers. By 1572 with his mother's brother he had set up a laboratory in an old abbey for the scientific study of astronomy and alchemy. It was here that he realized his first break: a new bright star that appeared in the constellation Cassiopeia. In a matter of months, Tycho became known throughout Europe for his precise determination of the position of the new star, now called a supernova. And not only did he prove that it was indeed a star, but also that it was positioned well outside the orbit of the Moon. Tycho can therefore be credited with demonstrating that the starry heavens were not immutable, as Aristotle had insisted.

A few years later, when King Frederick heard that the now-famous Tycho was planning to leave for the better astronomical facilities in Basel, he made Tycho an offer he couldn't refuse. The king would give Tycho an island in the middle of the sound near Copenhagen, called Hven, on which he could build his own observatory. The island would also be his personal fiefdom for perpetuity with all appropriate incomes.

Naturally, Tycho accepted the offer at once and moved to the island to supervise the construction of Uraniborg, his private palace and observatory. He wasted no time in starting his careful astronomical measurements. The indefatigable Dane not only supervised all the measurements with a ferociousness of a tyrant, but he also designed and constructed the massive instruments that were necessary to obtain the accuracy he demanded.

For the next twenty years, he and his slavish assistants carried out systematic observations of the sky, amassing a treasure of data on the motion of the planets and the position of the stars.

When his benefactor Frederick died suddenly in the middle of his historic tenure, Tycho managed to wrangle an agreement to secure the fiefdom for his heirs from the Regency Council. This was in spite of the illegitimate status of their birth, a result of Tycho's marriage to a commoner, which was not recognized by the nobility.

Eventually, in 1597, Tycho fell out with the new king, Frederick's son Christian IV, who was now coming of age. Apparently, the Lord of Uraniborg had been treating the island inhabitants harshly and neglecting the upkeep of a chapel where the young king's father and grandfather were interred. Difficulties with Christian continued and soon Europe's best-known astronomer fell into disgrace with the Danish court. He left Hven the next year, threatening to travel into voluntary exile.

Much to Tycho's surprise, the young king let him go, banishing him from his own homeland. The arrogant and determined astronomer was then forced to transport all his movable instruments and extensive entourage to Germany while earnestly seeking a new sponsor and a place to work.

While he was waiting on the continent, promoting his services to several European courts, Tycho received a startling book describing the structure of the universe from Johannes Kepler, a teacher of mathematics in a provincial Austrian town. Struck by Kepler's boldness and obvious mathematical prowess, Tycho immediately invited Kepler to join him as an assistant. However Kepler had just married a local widow in Graz, and a move to northern Germany was too far away for him to consider.

Nevertheless, Tycho and Kepler soon got together, almost by chance. When the Dane accepted an invitation as Imperial Mathematician to the court of Emperor Rudolf II in Prague, it brought him much closer to Kepler. What is more, the Lutheran schoolteacher in Catholic Styria was under increasingly severe persecution by the Counter Reformation, and was now in a position to take Tycho's offer more seriously. In 1599, after he refused to become a Catholic, Kepler was finally banished from his home. Frantically, he accepted Tycho's invitation. As the new century dawned, the two exiles in Bohemia were destined to start the golden age of astronomy.

They finally met on 3 February 1600 in Benatky Castle, 22 miles (35 kilometres) northeast of Prague, where Tycho was attempting to build another Uraniborg. Radically different backgrounds and temperaments meant that the meeting between the two men was tense. In Arthur Koestler's words, they '... met face-to-face, silver nose to scabby cheek ... opposite in every respect but one: the irritable choleric disposition which they shared.'

Once they started working Tycho decided to challenge the young upstart by assigning him with the research pertaining to the difficult orbit of Mars. Passing out only small bits of orbit information at a time, Kepler argued regularly with Tycho for several months. Aware that he was being patronized as a beginner, Kepler decided to leave. He pleaded with his former professors for an academic position at his old University in Tübingen but was refused. This left him no other option but to continue under the thumb of the overbearing Tycho, Imperial Mathematician to the Hapsburg Emperor.

Not the least of Kepler's frustration was Tycho's constant nagging him to use the strange cosmological system the Dane had devised in which most of the planets circle the Sun, but the Sun and Moon went around the stationary Earth. This was almost too much to bear for the headstrong German who had been convinced of the validity of the Copernican heliocentric system during his student days.

And then something unexpected happened. During a banquet on 13 October 1601 at the home of Petr Vok in Hradcany Palace, Tycho was indulging in his usual excessive eating and drinking. Feeling the urge to relieve himself, he waited in customary deference to his host who was still seated at the table. Suddenly, the corpulent nobleman collapsed on the banquet floor with a severe bladder ailment. By the time he returned home he could not urinate. After five days of sleepless agony caused by the acute uraemia, he found himself on his deathbed surrounded by his wife Kirsten and a few associates, including Kepler.

As fate would have it, Tycho's two closest colleagues, the trusted long-term assistant Christian Longomontanus and Tycho's son-in-law Franz Tengnagel, were both away from Prague. Realizing he was dying, the fifty-six-year-old Dane had no option but to promise the legacy of his astronomical treasure of measurements to the unsuspecting Kepler. He implored Kepler to use his hybrid system of the world – half ancient geocentric and half heliocentric – in calculating the planetary orbits: 'Don't let me have lived in vain', he cried, as he sank into unconsciousness. In spite of the Dane's pathetic plea, Kepler had no intentions of considering Tycho's world system. The cumbersome model could not possibly be reconciled with the German's idea of a physical force emanating from the Sun, which then moves the planets.

Kepler had long believed that if he could only get his hands on the volumes of accumulated data which Tycho had been hoarding all these years, he could prove that the heliocentric scheme and his own physical ideas were correct. He would be in luck. Following Tycho's elaborate burial ceremony in the grand Tyn Church on Prague's Old Town Square, the Emperor Rudolf II shocked the Habsburg Court by appointing the inexperienced Kepler as Imperial Mathematician. Rudolf charged him with completing the colossal task begun by Tycho of generating the complete elaborate mathematical tables for predicting the planetary positions for all time. With the autonomy of his prestigious position and access to all the measurements from Uraniborg for the very first time, Kepler now set out to revolutionize man's view of the universe. He hoped to develop a correct description of the motion of the planets, which he believed was caused by the force of the Sun. This was to be a new conceptual framework, a *physical* astronomy.

Johannes Kepler

In the quiet medieval town of Weil-der-Stadt in Germany's Black Forest, the impressive monument to Johannes Kepler dominates the main square. The statue shows Kepler sitting enthroned on an octagonal pedestal supported by relief carvings depicting his teacher, Michael Maestlin, his predecessor, Nicholaus Copernicus and his collaborator, Tycho Brahe.

Today hardly anyone visits Weil, remote as it is from any major cosmopolitan centre and removed from the main tourist trail. Unlike the beer halls of Heidelberg or the castles on the Rhine, there is no special reason to detour to this backwater – unless one knows the story of Kepler and his triumphant programme to overthrow ancient astronomy that began at the beginning of the seventeenth century. Kepler died at Regensburg in 1630 during the ravages of the Thirty Years' War and his grave was overrun by marauding troops. Its exact location is unknown – a far cry from the elaborate tombs of Galileo in Florence or Newton in London.

Fortunately, during the twentieth century a passionate group of *cognoscenti* in Weil-der-Stadt have acted to preserve

Kepler's legacy in the town and in 1930, at ceremonies in Weil commemorating the three hundredth anniversary of Kepler's death, the famous English astronomer Arthur Eddington travelled to Weil to pay his respects with a special address. Later, in 1971, an astute group of scholars from all over the world descended on the little town for a conference on Kepler's work to mark the four hundredth anniversary of his birth. Today, as appreciation for his work grows, tributes continue. In August 2009 in Prague an international conference called *Kepler's Heritage in the Space Age* was organized on the occasion of the four hundredth anniversary of the publication of Kepler's major work *Astronomia Nova*.

Strolling through the narrow streets of the provincial town today, one can imagine the scene on Christmas Eve, 1571, in the crowded little house on the corner of what is now called Keplerstrasse. If Johannes Kepler's diary recollections of his family can be believed, the Yuletide was accompanied by the heavy drinking of home-made brews and secret potions, violent arguing and praying as well as bizarre incantations.

In a corner of the warmest room was a young woman sipping one of her own infusions, concocted to keep away evil spirits from the foetus growing in her womb. Three days later, an infant boy would be born prematurely into a world where alchemy, astrology and witchcraft were still taken seriously. Johannes' mother, Katherina Guldenmann, the daughter of an innkeeper, and his father Heinrich Kepler, a mercenary soldier, lived with her parents and several other relatives in the small house in the centre of the provincial town.

The Kepler family was part of the Lutheran community in Wüttemberg, a Protestant enclave surrounded by the Catholic Habsburg Holy Roman Empire. Yet Luther did not impress Kepler's father, Heinrich. Records tell us he was a lustful soldier of fortune who could never settle down, repeatedly disrupting the Kepler household. When Johannes was only three, Heinrich joined a band of mercenaries fighting Protestant rebels in the Netherlands.

Johannes was a sickly and frail child who suffered from physical ailments throughout his adolescence. But his spiritual, philosophical and intellectual lives were as rich as his physique was poor. When the young man started his schooling, only one generation had passed since the death of Luther and the movement, which was to become known as the Counter Reformation begun. The Jesuits were in the process of developing an advanced educational system, which is still in evidence today. The clarion call had come from Ignatius Loyola in Spain: If the Muslims had used the sword for conversion, the Jesuits would use the classroom.

To compete with the Jesuits, the Lutheran Dukes of Wüttemberg had created an excellent school system of their own which they administered in the towns and villages of Germany. After winning a scholarship to one of these schools, Kepler thrived on a full academic programme in history, mathematics and science. The school curriculum also demanded intensive study of classical Latin with composition, grammar and public speaking. The boys in the Latin schools were forbidden to speak to each other in any other vernacular, thus becoming fluent in the ancient language. This superb education was cost-free, which was just as well, as the Keplers would not have spared a penny from their meagre resources for something as non-essential as schooling.

Given the opportunity to use his exceptional intellectual gifts, at the young age of eleven Kepler passed an important competitive examination in Stuttgart and gained a place in the Wüttemberg scholarship system. This assured his further education in the state's monastery school in Adelberg and, most importantly for the future, his higher education at the prestigious Tübingen University.

Having shown exceptional ability at an early age, he was given further special consideration by the authorities, who were only too pleased to encourage the development of a devout young Lutheran who may one day be preaching to the faithful. Taking his vocation as a theologian quite seriously, the young Kepler did not complain about the strict regime of the upper Latin school.

By the age of fifteen, he was already an intellectual who lived most of his life above his shoulders. The rest of his body seemed to be only a source of pain and distraction. The confusion and depression of adolescence turned into high-minded spirituality, expressing itself in the intense discipline of study. The incisive highbrow thinking that produced awe in his dealings with powerful men in his later years, not surprisingly caused resentment and even hatred in the minds of his adolescent classmates.

The 25th September 1588 was an important day for the fervent teenager, now hungry for academic success and acceptance by his school supervisors. During this fateful year, in which his father also abandoned the family forever, Kepler passed the baccalaureate examination for admission to the prestigious Tübingen University.

Tübingen University, and in particular its theological seminary, was one of the most renowned centres of German learning. Here Kepler thrived as an excellent university student, enthusiastically attending lectures in ethics, dialectics and rhetoric as well as Greek and Hebrew, which were necessary to decipher the original biblical texts. In addition, there was astronomy and physics, which would later become so important to the unsuspecting seminarian.

In time, some of his teachers noticed a more developed interest in astronomy and mathematics than would be expected of a theology student. Of these teachers, Michael Maestlin was the most important. An esteemed member of the Tübingen faculty at the time of Kepler's arrival, Magister Maestlin was twenty years older than his young charge, and one of the most capable astronomers of his time. He taught Ptolemy's system of the world in his lectures but also discussed the Copernican system. He was perhaps the only academic in Europe to give serious attention to heliocentrism and deserves a great deal of credit for introducing Kepler to Copernicus. It has been reported that the professor met secretly with a small group to discuss the new system, fearing the repercussions of any public espousal of the Copernican system.

Maestlin had a curious attitude concerning the two world systems. For example, his personal copy of Copernicus' *De Revolutionibus* is the most heavily annotated copy in existence. Yet in 1582 he wrote a major work on Ptolemaic Astronomy, *Epitome Astronomiae*, in which there was no mention at all of the heliocentric system. Even years after Kepler had established the validity of the Copernican model with his work on the orbit of Mars, Maestlin still had not openly advocated the 'new' system. Nevertheless the ever-generous Kepler acknowledged his great debt to Maestlin from his university days in many of his letters and books:

> Already in Tübingen when I followed attentively the instruction of the famous Magister Michael Maestlin, I perceived how clumsy in many respects is the hitherto customary notion of the structure of the universe. Hence I was so very delighted by Copernicus, who my teacher very often mentioned in his lectures.

In August 1591, Kepler received his MA from the University and entered the three-year theology course based in The Stift, the most prestigious seminary for aspiring Lutheran theologians in Europe. Reports on Kepler's behaviour in these years indicate that he focused solely on the pulpit and the subject of astronomy had little importance to him. Kepler himself later wrote about his courses in science and mathematics – the prescribed studies – but said nothing to indicate any particular inclination for astronomy.

From Magister Maestlin's initial lectures on the various world systems, Kepler seemed to instinctively glean that Copernicus was right. The beauty and simplicity of the heliocentric model appealed to his aesthetic sense of how God had made the universe.

Twenty years earlier, the world's most accurate astronomical observer, Tycho Brahe, had rejected Copernicanism because he could not detect any stellar parallax, which should result from viewing the stars from a moving Earth. Kepler's opponents at the university did not need such proof – the idea of humanity's home in the universe flying through space at the speed required by the heliocentric model at 67,000 mph (108 kph) or about 1,000 times faster than the speed of an auto on a highway, was ridiculous to most. Still, Kepler argued with his cynical fellow students and demonstrated the consequences of such motion using the practical astronomy he had learned from Maestlin.

Religious conflicts haunted Kepler throughout his life, and this had a very particular impact on his passion for astronomy. He was sure Copernicus' system was correct; however, at Tübingen he was left with no other choice than to support the Ptolemaic geocentric model adopted by the Lutherans. He had not been censored within the safe confines of Wüttemberg and had not yet faced the animosity of the Catholic scourge of the Counter Reformation that was otherwise rampant in other parts of the Holy Roman Empire. This movement would attempt to dictate not only what he could believe about the motion of the planets, but also how he could practise his faith. To Kepler, whose life was built around his relationship to his creator, such dictates would prove to be intolerable restrictions on his personal freedom.

Reading the diaries and letters which Kepler wrote during his years as a seminarian, it would seem that in addition to science, he had a complex and well-reasoned opinion on just about everything. In the later years of his life – and centuries later, by the historians who assessed him – this penchant would bring both admirers and detractors. But Kepler was not pretentious; he simply could not ignore any challenge to his formidable intellect, particularly in matters of religion and science.

If intellectually Kepler was clearly blessed then physically he seemed to carry a curse. In addition to the catalogue of ills that fill his personal horoscope of 1597, he also complained of terrible haemorrhoids, which often required him to work standing up rather than sitting at the table. According to his diary, he only chanced a bath once in his life. Against his better judgment and after constant nagging from his wife, he reluctantly immersed himself in a hot tub when he was about thirty years old. 'Its heat affected me and constricted my bowels', he wrote.

Many historians have noted on Kepler's remark that as a child he never felt any particular inclination towards astronomy. However, recollections made in his mid-twenties give a fascinating record of what did impress Kepler as a child, namely seeing a comet in the sky at the tender age of six years. The schoolboy's mother had roused him from sleep and led him to the slope of a hill to see the rare 'star with the hairy tail' spread across the sky. Kepler remembered this incident with clarity and wonder, even if during his university years he may not have been aware of the intensity of his interest in astronomy.

A New Direction

In the spring of 1594 the earnest seminarian was looking forward to becoming a Lutheran minister. However, Georgius Stadius, the mathematics teacher at the Lutheran Stiftsschule in the provincial Austrian town of Graz in Catholic Styria, died suddenly. As established procedures dictated, the Lutheran authorities in Graz immediately sent a request to Tübingen University for a teaching replacement. In one of the most bizarre shifts among many in Kepler fortunes, he was selected by the University senate to fill the post.

After much soul-searching and tense discussions with the university authorities, the theology student accepted the teaching position as a temporary assignment, no doubt with humble gratitude for the entirely free education he had received under the patronage of the Lutheran Dukes. Kepler collected his belongings, his books and his pride and set off for Graz. Could he survive in the outside world, completely responsible for himself for the first time in his life? He would be travelling to an unfamiliar land, one dominated by the Catholic Habsburgs.

In Graz, the strong-headed, opinionated teacher would face a new adversary, the Roman Catholic Church. Here in the late part of the century, the pressures of the Counter Reformation were intensifying as Catholic rulers schemed to rid their lands of heretics and to nullify the concessions that Emperor Maximilian had made to the Protestants many years before. Controversy began to swirl amongst the Lutherans in Graz as the Counter Reformation gathered momentum.

Given the tensions between Catholics and Protestants during the final decade of the sixteenth century, Graz would seem the last place to send a young seminarian who was inflexible and outspoken in his religious beliefs. Nevertheless, on 11 April 1594 (according to the Gregorian calendar) Kepler arrived in the provincial town, entering a religious and political atmosphere quite different from the Lutheran enclave of Tübingen.

Just over a year after his arrival, a remarkable incident occurred while he was lecturing to his astronomy class. He made a discovery, which he believed revealed how God had made the universe. This vision would affect everything he did in astronomy from that day forth. On that fateful morning in July 1595, Kepler was introducing his small class to the unusual notion that the outer planets Jupiter and Saturn produce a conjunction every twenty years. That is to say that the two planets, one behind the other as viewed from the Earth, appear to be at the same point in the sky.

The conjunction first appeared in the constellation Aries and then appeared in Sagittarius; next in Leo and then back to Aries, albeit not in the same exact location as the original conjecture. Each conjunction was seen to move about 9 degrees forward in the direction of the movement of the zodiac. If straight lines are drawn connecting these points, an interesting symmetric diagram is generated.

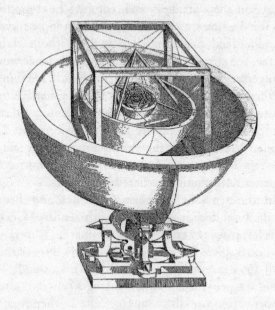

Kepler's model of the spacing between planetary orbits, 1595.

At a time when there was no distinction between astrology and astronomy, such recurring conjunctions of the two major planets, eight zodiac signs displaced from each other, was note-worthy. Kepler, who had some respect for astrology, proceeded to draw straight lines connecting the successive points on the zodiac circle where the conjunctions were observed over many years. Surprisingly, this formed a series of equilateral trian-gles, each rotated 9 degrees from the other around the zodiac. Stepping back, he observed that an inscribed circle had been generated by the outline of the points where the triangles intersected. The symmetry of the final diagram could be repre-sented by a single equilateral triangle inscribed within the two circles. Suddenly, he noticed the ratio that the diameters of the circles were close to the ratio of the diameters of the orbits of the two planets as published by Copernicus. Inspired by his discovery, he imagined that by inserting different geometric figures between the planetary shells he might explain the spacing between the orbits of the planets.

He immediately considered other geometric figures bearing high symmetry relative to their sides. Further calculations showed that simple two-dimensional figures like a triangle and a square did not give very good agreement with relation to the spacing between planets. So he considered three-dimensional solids. Is not the universe three-dimensional, he asked, reassuring himself. The triangle therefore became a tetrahedron; the square became a cube, etc.

He also interchanged the various geometric figures in order to get the best agreement. He replaced the original triangle, the two-dimensional figure between Jupiter and Saturn, with a cube, a three dimensional figure, which gave a better result. Proceeding in this manner, he used only the regular polyhedra known to Euclid, Pythagoras and Plato, the five so-called 'perfect' solids (a perfect solid is one whose faces are all identical). As there were only five interplanetary spacings, his scheme could thus only accommodate six planets, no more or less. However, at that time, there *were* only six planets known to exist in the universe, excluding the Moon, which in the Copernican scheme is a satellite of the Earth, and not a planet.

Thus, his model accounted for the number of planets thought to exist at that time, and he was hopeful that it would also predict the spacing between the orbital shells in some agreement with contemporary values. In any case, Kepler believed he had discovered God's blueprint. This remarkable conclusion he described in his first scientific publication, *Mysterium Cosmographicum* (The Secret of the Universe), in 1596. In his attempt to understand the structure of the universe, it is arguable that this may have marked the beginning of Kepler's transition from astrology to astronomy.

Had the schoolteacher seen in a flash how God created the universe? If so, this revelation suggested to him that the secret had been revealed for a purpose, to help him serve his god as an astronomer if not as a clergyman. The resulting model of creation, developed from the inspiration of that summer morning and full of spiritual and mystic overtones, would haunt him for the rest of his life. In spite of much evidence to the contrary, he would never – even on his death thirty-five years later – accept that his model was wrong. But since Kepler was guided by this first concept of the universe to other, correct, astronomical discoveries, the *Mysterium Cosmographicum* is very important to the history of modern science.

Kepler, the former seminarian, was thus proposing that he knew how God had built the universe. This idea can be taken as either utterly profound or totally ridiculous.

He found that optimum agreement with the spacing between planets as published by Copernicus would be obtained if the perfect solids were placed in the following order between the planetary spherical shells:

- Between Saturn and Jupiter – a cube (6 sides)
- Between Jupiter and Mars – a tetrahedron (4 sides)
- Between Mars and the Earth – a dodecahedron (12 sides)
- Between Earth and Venus – an icosahedron (20 sides)
- Between Venus and Mercury – an octahedron (8 sides)

Taking the radius of the Earth's orbit as unity, the relative radii of the other planetary orbits determined by Kepler compare quite favourably to Copernicus' values and to modern values, as the table below demonstrates:

Planet	Copernicus	Kepler	Modern Values
Mercury	0.395	0.429	0.387
Venus	0.719	0.762	0.723
Earth	1.000	1.000	1.000
Mars	1.512	1.440	1.524
Jupiter	5.219	5.261	5.203
Saturn	9.321	9.163	9.539

Table 1: Radii of Planetary Orbits

Everything seemed to fit within about 5–10 per cent, some-what astounding for a first attempt at a theory of the entire known universe. If a twenty-first century research astronomer investigating a parameter as uncertain as the relative size of the planetary orbits lived during Kepler's time, such agreement would be considered remarkable.

Encouraged by his mentor Maestlin's surprising interest in his discovery, Kepler asked for a short leave from the school authorities in Graz to travel to his homeland in order to confer with Maestlin regarding his creation model of the nested solids. With more than his usual abundance of excitement, he also privately discussed the completion and publication of his book with his former teacher.

Both Maestlin and the University Senate gave their support, albeit with some comments regarding the book. They suggested that he present the ideas in a more simple and clear language and reminded him not to assume that everyone was already a Copernican and totally familiar with the complexities of a Sun-centred system. Kepler accepted their advice with enthu-siasm. In the first chapter in particular, he presented a clear and detailed exposition of the advantages of the heliocentric scheme over the old-fashioned epicycle wheels of Ptolemy. The elucidation of Copernicus' theory found in this first chap-ter alone would mark him as the most important advocate of the new system during the Renaissance.

The year 1595 was a momentous one for Kepler. In early December Kepler had become attracted to a local young woman, Barbara Mueller, who had been presented to him as an ideal match. The first-born daughter of a wealthy mill owner, she was described as being 'pretty and plump' and was twenty-three years old when she met Kepler. She had already been married twice, with both of her husbands dying after a few years of marriage. Her father, Jobst Mueller was sceptical of a third marriage, particularly to an astronomy teacher with a meagre salary. But Jobst had few alternatives, and so his daughter and Kepler were married on 27 April 1597 with the great splendour that befit the custom of the time. Thereafter, Kepler moved into the comfortable home of the mill-owner's daughter in the Stempfergasse in the centre of Graz. This increased his income by a significant amount, as he was no longer required to pay rent for accommodation at the Stiftschule.

Shortly after, in the spring of 1597, Kepler got his hands on printed copies of *Mysterium Cosmographicum,* the final version of his book. He was then quickly thrust into the front ranks of European astronomers after sending fifty copies to Maestlin to distribute in Tübingen and individual copies to the leading scholars in Europe, including Galileo and Tycho. His reputation spread.

Sadly *Mysterium Cosmographicum* is full of so many radical ideas about the heavens and written in such an unusual style it was easy for his contemporaries to ignore, and even ridicule Kepler's work. *Mysterium Cosmographicum* is today no more considered a fundamental breakthrough by most contemporary historians than it was upon its first appearance in the sixteenth century.

The Danish science historian J.L.E. Dreyer considers *Mysterium*'s first chapter, which Kepler wrote to satisfy the Tübingen University's request to explain the Copernican system more clearly, a major piece of elucidation. There in a simple, popular style are set down the reasons for abandoning Ptolemy in favour of Copernicus. Over three hundred years after its appearance, Dreyer stated in his classic treatise *The History of the Planetary Systems* that it is difficult to see how anyone could read this first chapter and still remain an adherent of the Ptolemaic system.

So what exactly was in Kepler's book? Kepler was suggesting not only how the Creator had made the universe, that is with the 'perfect solids', but also how the Almighty had set the universe in motion. He was offering physical explanations for celestial phenomena. Starting from the obvious fact that planets further from the Sun move more slowly, he assumed that the Sun was not just a lamp in the centre of the universe giving off heat and light. He postulated that the Sun also provided a driving force that diminished with distance and that it also moved the planets in their orbit. This was the first time in history any physical role was attributed to the Sun.

But what was the nature of this force from the Sun? This proposed interaction between the Sun and the planets was an attempt to modify the thinking that had prevailed for fifteen centuries. Previously the heavens and the Earth were thought to be completely different, separated into their own distinct realms and composed of different substances: one 'heavenly' and perfect, the other 'Earthy' and corruptible. In order to study the sky, astronomers had been content to construct mathematical models – mere calculating devices – that could predict the position of the planets. That's why for centuries, astronomy was considered a branch of mathematics. But these ideas were beginning to be challenged.

Tycho had questioned this dichotomy when he discovered the Nova in 1572 and Kepler was now causing more doubt. Though Kepler's geometric solar system – constructed of perfect Platonic solids – made many scientists and historians (both then and now) wary of his model, his book is important for its clear and convincing justification of the Copernican scheme. Furthermore, Kepler's insistence on a physical role for the Sun in moving the planets made *Mysterium* one of the seminal books in the history of science. Up until this time, astronomers believed the Sun merely provided light and heat for the Earth. Now Kepler was implying that the Sun caused the actual *motion* of the Earth, as well as the other planets. This was essentially the beginning of enquiry into the physics of the skies.

Kepler never lost faith in his geometric model, and his incessant strive to prove it led him to other major discoveries throughout his career. In his last important work *Harmonice mundi* (Harmony of the World), Kepler still claimed that the direction of his life, his studies and his work had in fact originated from this one book, *Mysterium Cosmographicum*.

When Kepler did begin to receive responses and opinions on the *Mysterium*, he found himself gaining in confidence, especially from the endorsement of his own mentor, the much-respected Maestlin. But there were two other important and well-known scientists whose responses would cause him great anxiety: the Italian Galileo and the Dane, Tycho Brahe, who would significantly influence Kepler's future career.

Unbeknownst to Kepler, Tycho had already left his island observatory in a dispute with the young Danish king, Christian IV. He and his family, as well as his astronomical assistants, were on their way to Prague with a commission as Imperial Mathematician to the Holy Roman Emperor, Rudolf II. It was only here that Tycho became aware of the *Mysterium*.

But Kepler had other problems. The Catholic Counter Reformation was gaining the upper hand in Styria and by September 1598 most Lutheran schools, including Kepler's Stiftschule, had been closed. The hapless mathematics teacher was now unemployed and unsure how to proceed. He did have some good fortune, mainly due to his reputation as an impressive intellect and the successes he gained from his astrological predictions. Government authorities allowed him – as one of only a few Lutherans – to stay in the city and the school continued to pay him his modest salary. It is clear that the inspectors did not want to limit his great scientific career and even encouraged him to use his philosophical leisure for the advancement of mathematical science. In the end, he saw the situation as an opportunity to return to his contemplation on the harmony of the world. Yet knowing that the Catholics and the Jesuits had more oppressive tactics planned, Kepler could not fully relax. He knew he would eventually have to leave.

By now, Kepler knew Tycho was in Prague as he had received word from a friend at court that the Emperor Rudolf had brought the famous Dane to the Habsburg court. This raised Kepler's hopes of a position with Tycho in Rudolf's Prague. It would also provide a double fulfilment: in a cosmopolitan city ruled by a liberal and tolerant ruler, Kepler would find a haven from the extreme persecution conducted by the Styrian (southern Austrian) Catholics. Secondly, he would be able to obtain the observations he needed to resolve the uncertainties in his astronomical studies. In addition, he found out that Tycho was being paid a magnificent salary of 3,000 gulden, fifteen times that which Kepler received from the school officials in Graz.

It was another stroke of luck when, in early 1600, Johann Friedrich Hoffman, a member of the Styrian diet and a counsellor to Emperor Rudolf, was planning to visit the Prague court. Already impressed with Kepler for his work as district mathematician, he offered to take Kepler with him to Prague and personally introduce him to the new Imperial Mathematician, with whom he had previously been in contact. They left for the Bohemian capital on the first day of the new century, 1 January 1600. Upon arriving in Prague, Kepler stayed as Hoffman's guest where he immediately received an exciting welcoming letter from Tycho:

> You will not come so much as a guest but as a very welcome friend and highly desirable participant and companion in our observations of the heavens.

Meanwhile, technical help for Tycho had arrived from his native Denmark. Though his facilities were certainly not ready, his most respected assistant from Hven, Longomontanus, was in Prague and ready to work.

With Kepler on his way and his experienced associate from Hven already in Bohemia, Tycho was itching to begin a new 'golden age' in his adopted homeland of Bohemia under the patronage of the Holy Roman Emperor. He decided to push ahead with the most complex problem, determining the shape of the orbit of Mars from previously unused measurements taken on Hven.

When Longomontanus arrived he was therefore immediately assigned the problem of calculating the orbit of Mars. He was instructed to use Tycho's data as well as the Dane's awkward hybrid system of the geocentric universe. It was a tough assignment. Mars had never revealed its secrets to previous investigators but it seemed it would be different this time. Working in Prague with the Uraniborg observations, Longomontanus quickly accounted for the longitude positions of Mars' entire orbit to an accuracy of two minutes of arc. This might seem impressive were it not for the limited parameters of his procedure. First of all, he used only Tycho's observations taken at opposition, that is when Mars, the Earth and the Sun were aligned. This simplified his analysis.

Secondly, he paid no attention to the latitude variation of the planet, assuming that positions above and below the ecliptic were observational errors. This highlighted the assumption that all the planets moved almost exactly along the same path as the Sun and that latitude values were therefore always small. Using this logic, the only coordinate to worry about was the longitude. So a full description of a heavenly body's position on the celestial sphere required that only the planetary longitude be known. Identifying the planetary longitude meant measuring the angular position around the ecliptic, taken from some arbitrary point called the Vernal Equinox, which was a kind of Prime Meridian of the Sky. (To be exact, this is the point in the sky where the Sun annually crosses the ecliptic on its northerly path from the southern hemisphere of the celestial sphere to the northern hemisphere.)

However, it was also important to understand the latitude, the angular distance above or below the ecliptic, which Longomontanus chose to ignore. The latitude has a much easier reference point or zero value – namely the ecliptic itself. Prediction of the longitude is a far more important and difficult task. Indeed, accounting for the apparent non-uniform path of Mars around the ecliptic, especially considering its retrograde motion in the sky, had first compelled the ancient astronomers like Ptolemy to introduce equants and epicycles into their models of the heavens.

A single theoretical model that could predict the longitude *and* latitude for each planet would therefore be preferred. Such a model could be used to generate ephemeris tables that gave the two angular co-ordinates of each planet for all time, past and future. Navigators at sea could observe any two planets in the sky, find their angular positions from the tables and by triangulation, then determine the position of the observer on Earth.

This task was at the centre of Tycho's service as Imperial Mathematician to the Emperor – especially given that Emperor Rudolf was constantly seeking the most reliable astrological prognostications. Consequently, his support for Tycho was boundless. Thus, in Prague at the beginning of the seventeenth century, thousands of the most accurate observations that Tycho had accumulated at Uraniborg stood ready to test any theoretical model that could predict the longitude and latitude of Mars or any of the other planets.

In order to complete his task, Tycho needed a very clever mathematician who could analyse the observations. Someone had to derive order from the endless numbers recorded in the giant leather volumes that Tycho had transported across Europe from Hven. Longomontanus didn't seem to be up to the task.

Tycho's decision to take the problem of resolving Mars' orbit away from Longomontonus and re-assign it to Kepler was one of the great moments in the history of astronomy. Using the astronomical observations carried out by the Dane, Kepler would eventually solve the orbit by discovering a single general law of elliptical orbits for all the planets. It is recorded that Kepler made a wager with Longomontanus that he would crack the problem in a week. The stakes were not known, but the Danish astronomer won easily. It took Kepler five years to find the solution.

For his own part, Kepler had not planned to focus on an individual planet, and was still interested in obtaining better values for the orbital eccentricities of all the planets. He wished to complete his own project on the 'harmony of the world', one which had begun with the *Mysterium Cosmographicum* in 1597. He described the situation in a letter:

I would have completed my research on the harmonies of the world if Tycho's astronomy had not fascinated me so much that I almost went out of my mind . . . One of the most important reasons for my visit to Tycho was the desire to learn from him more correct figures for the eccentricities in order to examine my Mysterium and Harmony for comparison. For these speculations a priori, must not conflict with the experimental evidence; moreover they must be in accordance with it. But Tycho did not give me the chance to share his practical knowledge except in conversations during meals . . . today something about the apogee, tomorrow something about the nodes of another planet. But when he saw that I had a daring mind, he decided perhaps the best way to deal with me was to give me my way with the observations of a single planet, namely Mars.

Kepler immediately applied his own radical ideas to his new assignment. Not only did he ignore the various geocentric models in use by Tycho and his acolytes, but he also introduced a totally new role for the Sun in the heliocentric scheme.

Kepler started with Copernicus' basic idea that the Earth spun about its axis while moving in an orbit about the Sun. As with everyone else before him, he first tried to plug the data into a circular orbit, adjusting the position of the Sun to get the best fit for a possible orbit. Unlike the others however, he had rejected the use of epicycles, which to him made no physical sense; he found the concept of motion around an empty centre (that is, with no object within to produce the curved motion) unacceptable. If this process of curve fitting had been easy, he would have succeeded immediately and won the bet he had made with Longomontanus.

However, the calculations were very complicated. He had to translate Tycho's determination of the angle made by the planet with respect to the fixed stars as seen from the Earth into another geometric system. And this other geometric system could not be more different in its consideration of a space defined with respect to a fixed Sun about which the Earth was moving. He made some seventy separate attempts, each of which involved long and time-consuming calculations that slightly adjusted the position of the Sun to optimize a fit.

Finally, he managed to produce a curve that agreed fairly well with the observations. This he named his 'vicarious theory'. However, when he included additional measurements from some of Tycho's other records, he was disheartened to see the agreement shattered. After all this hard work, he was about to give up even though the difference was very small, only 8 minutes of arc or 2/15ths of one degree.

Yet Kepler had unshakable faith in the accuracy of Tycho's measurements. These were, after all, accurate to within only 2 minutes of arc and so he accepted the observations over the mathematical curves he had constructed. Kepler decided to throw out his vicarious theory, and with it all of the other ancient theories of planetary orbits to start anew. In so doing, Kepler made the right decision – for Tycho's measurements were correct. In a spectacular testimonial to the spirit of the Renaissance, he wrote in his journal:

> If I had believed that I could ignore these 8 arc-minutes, I would have patched up my hypothesis accordingly. But, since it was not permissible to ignore, those 8 arc-minutes pointed to a complete reformation in astronomy. Upon this 8-minute discrepancy, I will yet built a new theory of the universe.

Kepler began once again, carrying out agonizing analysis on Tycho's measurements of Mars by using an ingenious scheme for finding the positional points of the orbit. He knew the period of Mars' orbit around the Sun to be 687 days. The planet would thus be in the same position every 687 days. He could then take sight lines from the Earth against the fixed stars in order to locate points on the Mars orbit.

Using this method, he found he could not fit all of the measure positions to the *uniform* motion of the planet in a modified Copernican orbit that was perfectly circular. Again he was faced with a small but persistent discrepancy of about 8 minutes of arc, only a fraction of the degree. Kepler knew Tycho's accuracy was such that only an error margin of much less than that, perhaps only 2 minutes of arc, would have been tolerated. On this basis, he decided to discard the entire ancient system of cosmology and start anew. With great conviction he wrote:

After the divine goodness had given us in Tycho Brahe so careful an observer, that from his observations the error of calculation amounting to eight minutes betrayed itself, it is seemly that we recognize and utilize in thankful manner of this good deed of the Gods, we should take the pains and search out at last the true form of the heavenly motions.

Kepler's Laws of Planetary Motion

So Kepler began again, changing the speed of the planet as it moved in its orbit around the Sun and thus discarding another ancient and cherished belief. As a guide, he used an imaginary spoke on a wheel connecting the Sun to the planet. In experimenting with this arrangement he made his first great discovery. He found a surprising fact: This imaginary spoke, driven by the data of Tycho, *swept through equal areas in equal time periods*.

This came to be known as Kepler's second law of planetary motion though it was discovered before the so-called first law. With this discovery of the *area law*, Kepler finally abandoned once and for all his attempt to simulate planetary motion with combinations of circles as the ancients had done and began to experiment with 'ovals' as orbits for Mars.

After more calculations, he achieved his most important result, the *law of ellipses*, now known as his first law of planetary motion. Using the law of areas he was able to show that all planets, beginning with Mars but extending to all the others, moved on elliptical orbits with the Sun at one focus of the ellipse. After years of effort, he was elated with this result, and saw the hand of God in the ellipse, which to a mathematician like Kepler was as 'perfect' as a circle.

Thus it was that Kepler found the simple curve – known even to Euclid – which fit the observed motions of all the planets. He was able to confirm that the planets moved on elliptical paths in such a way as to sweep out equal areas in equal times. Kepler's successful plotting of the orbit of Mars was to become the most important event in the explanation of motion in the heavens at the dawn of the seventeenth century. However, it was not until 1605 that Kepler was able to announce his new discovery.

Over the next decade Kepler went on to formulate his third law of planetary motion, which he called the law of periods, which related the periods of planets to the sizes of their orbits. In essence, this is really a relationship that describes the structure of the solar system. Initially conceived in his first book on cosmology in 1596, it did not appear until twenty-three years later after he had left Prague when he published his *Harmony of the World* in Linz in 1619. As he explains:

> ... if you want the exact moment in time, the correct form of the law was conceived mentally on 8th March in this year 1618, but submitted to calculation in an unlucky way, and therefore rejected as false, and finally returning on the 15th of May and adopting a new line of attack, stormed the darkness of my mind. So strong was the support from the combination of my labour of seventeen years on the observations of Brahe and the present study, which conspired together, that at first I believed I was dreaming, and assuming my conclusion among my basic premises. But it is absolutely certain and exact that the proportion between the periodic times of any two planets is precisely the sesquialterate proportion of their mean distances ...

Since the time of the Greeks it had been known that the planets with the largest orbits took the longest time to complete a cycle. Kepler was always hoping to discover a more quantitative connection and finally, after many trials over several years, he found the regularity between time and distance he was searching for, *the ratio of the average distance of each planet from the Sun, cubed, to the period of revolution for each planet, squared, is the same constant value*. This was something like an exercise in numerology, but once he happened upon this ratio, the regularity was striking, fulfilling his predilection for hidden resonances in God's creation. The regularity in the value of this ratio suggested that there must be an underlying law of nature yet to be discovered. (Newton found this law in 1687, the law of universal gravitation).

Planet	R (in AU)	T (in sec) x 10^6	R^3/T^2 x 10^{13}
Mercury	0.387	7.60	0.998
Venus	0.723	1.94	0.995
Earth	1.000	3.16	1.000
Mars	1.523	5.94	0.996
Jupiter	5.202	3.74	0.994
Saturn	9.554	9.30	0.990
Uranus	19.218	2.66	1.000
Neptune	30.109	5.20	0.990

Confirmation of Kepler's 3rd Law

The table above shows the ratio of the average distance cubed to the period squared for the planets, using modern values of one-half of the major axis of the ellipse in Astronomical Units and the period in seconds.

This was a momentous result. Twenty years of observations and thousands of measurements were condensed into a simple system of curves and rules. Thereafter, anyone trying to construct a system of the heavens must reproduce these three laws in order that their theory gives the correct motion of the planets. Kepler wrote of his own result in his journal:

> What sixteen years ago I urged as a thing to be sought, that for which I joined Tycho Brahe . . . at last I have brought to light and recognize its truth beyond my fondest expectations . . . The die is cast, the book is written, to be read either now or by posterity, I care not which – may well wait a century for a reader as God has waited six thousand years for an observer.

Astronomia Nova
Kepler's most important work *Astronomia Nova*, published in 1609, contains the results of the astronomer's nine-year long investigation of Mars. The full English title of his work is *New Astronomy, Based upon Causes, or Celestial Physics, Treated*

*by means of Commentaries on the Motions of the Star Mars,
from the Observations of Tycho Brahe* – quite a mouthful. This
work represented a real departure point for modern astronomy.
Where previous astronomers had relied on geometric models
to explain the observed positions of the planets, Kepler sought
for and discovered physical causes for planetary motion. What
is more, he described the motion with algebraic laws. In 1543
Copernicus proposed that the Earth and other planets orbit the
Sun, but Kepler was the first astronomer to prove this conjecture with rigorous scientific arguments.

At over 650 pages in English translation, *Astronomia Nova* is
a lengthy and difficult work from the perspective of a modern
reader, much as it was in 1609. Kepler walks his readers step-by-
step through his process of discovery, including the false starts,
the dead ends and the mistakes. In the process of so doing he
was trying to dispel any impression of 'cultivating novelty', or
doing anything original with Tycho's data. But there were other
reasons for Kepler to employ this particular narrative technique.

We know now from recent research why is it so difficult
to follow the sequence of arguments in *Astronomia Nova* by
which Kepler progressed from a modified Copernican model
based on uniform circular motion to the non-uniform elliptical motion and the discovery of his first two laws of planetary
motion. He wished to demonstrate to his critics at the time,
and in particular to Tycho's supporters and heirs, that he had
tried at the outset to use Brahe's hybrid geocentric/heliocentric
model, as per the conditions that Tycho outlined in using the
legacy of the Hven measurements.

But this was a deceit. Cleverly disguising his unwavering
acceptance of the Copernican scheme, Kepler had tried to
show in the final manuscript how he had attempted to align
Tycho's data with the world systems of Ptolemy and Tycho
before considering the heliocentric model. He also pretended
that he had simply stumbled across the discovery of his
planetary laws. This was a ploy to write his way *around* the
obstinate resistance of Tycho's heirs and supporters, most

notably Tycho's son-in-law, Junker Tengnagel and the faithful assistant Longomontanus, in order that he be able to proceed with the publishing of his conclusions on the perplexing orbit of the planet Mars.

Clearly, this has led to criticism of *Astronomia Nova,* which has divided scholars from Galileo – a contemporary of Kepler – up to the present day. Kepler's main biographer Max Caspar, writing in German in 1937, for example, thought that Kepler's confusing journey had been an interesting story of discovery in which he tried to tell events as they happened. Later, Arthur Koestler, working mainly from the translation of Caspar's German text in a widely read popular account titled *The Sleepwalkers* (1959), concluded that Kepler was simply unable to write in an orderly manner. Yet other scholars tried desperately to string together the logical sequence in *Astronomia Nova.* The American scholar Curtis Wilson completed an article in the *Scientific American* on 'How Kepler Discovered His Laws of Motion'. Like him, many other scholars were puzzled by the conundrum in the structure of Kepler's writing.

However, thanks to the careful research of James R. Voelkel, whose doctoral thesis exposed Kepler's intentionally rhetorical writing in *Astronomia Nova,* we now know the difficult style and approach in the book was probably done purposely to mislead readers, particularly Tengnagel and Longomontanus. There is no doubt amongst experts today that the threat of editorial interference was responsible for *Astronomia Nova*'s unusual narrative form. Kepler sought to nullify objections to its publication by showing that all other possible approaches had been tried and that the heliocentric scheme with elliptical orbits was the only one possible.

Kepler's *Astronomia Nova* ranks with Copernicus' *De Revolutionibus,* Galileo's *Two World Systems* and Newton's *Principia* as the seminal works of the scientific revolution. In fact, a deeper appreciation of Kepler's role in this development is now evident in the scholarship of contemporary historians of science.

7

GALILEO GALILEI: THE
ITALIAN CONNECTION

Since the period over which the scientific revolution took place
was only 144 years – from Copernicus' *Revolutionibus* in 1543
to Newton's *Principia* in 1687 – it is not surprising that many
of the great discoveries came as a result of resource collaboration and data sharing. We have already seen how Kepler's
work was entwined with that of Tycho Brahe, resulting in their
fruitful collaboration in Rudolf II's court in Prague. A wonderful symbiosis existed between these two men; one was a great
experimentalist, the other a great mathematician. On the other
hand, Galileo – the astronomer of Pisa – had no collaborators,
at least none that he ever acknowledged.

In many ways Galileo was a blend of Tycho and Kepler in
that he contributed to both observation and theory. Like Tycho
he knew the importance of careful experimentation. He also
designed his own apparatus and single-handedly carried out
many of his own measurements; thought of himself as a rough-
edged aristocrat; married a common-law wife (probably for the
same reasons as Tycho, that is that no noblewoman would have
him); and possessed the same irascible and aggressive manner
in his dealings with others as did the 'Lord of Uranibourg'. Yet
Galileo, unlike Tycho, also possessed the genius of mathematical

reasoning that Kepler possessed. Like Kepler he understood the application of theory in the physical world and was intent on developing laws that could be tested – either in a laboratory or in the heavens – by experiments that he himself designed.

Kepler's main theoretical contribution to planetary theory was to devise laws based on Tycho's observations. Not withstanding his remarkable discoveries using the telescope, Galileo did likewise. He was more than an astronomer, and focused his attention on new ideas, which comprised elements of physics. These include such concepts as time and distance, velocity and acceleration, force and matter. In his studies and hypotheses of free-falling bodies and relative motion, his methods became the prototype for future science.

Historians of science admire Galileo for asking crucial questions about motion that no one else seemed to think were important. The answers to these questions jarred dramatically with the accepted view espoused by scholastic Aristotelians who spoke of essences, philosophies, metaphysics and causes, while Galileo discussed proofs and demonstrations. Furthermore, in keeping with Kepler's new method of applying mathematics to physical reality, Galileo's studies were innovative in his insistence on applying concepts and mathematics to strictly observed facts that could be tested.

The approach of the Italian was therefore ultra-pragmatic and totally new. In one experiment, a preliminary to understanding free-fall, he observed the motion of a ball accelerating down an incline which was set at different angles. He timed the ball's motion with a crude water clock. This led to a clear concept of acceleration, the rate at which the speed of an object changed. Galileo saw the fundamental importance of understanding the exact details of phenomenon such as acceleration without being distracted by the philosophical problems that most academics believed still needed to be solved.

Early Life
Galileo was born in 1564 at Pisa, an important city on the River Arno in Tuscany, Italy. He was the eldest of six children

born to a famous musician, Vincenzo Galilei. When Galileo was quite young, his father moved the family to Florence and enrolled his eldest son in a monastery at Vallombrosa near the city.

As a young man Galileo was expected to become a priest, but this vocation did not suit him so he left the monastery and registered for a medical degree at the University of Pisa. However, he did not complete the degree, switching to the study of mathematics instead. This subject suited him and as his knowledge grew so did his interest in scientific problems such as physical measurements and applied mathematics.

However, it was during the years that he was in Pisa that Galileo became fascinated with the work of the ancient Syracuse scholar Archimedes. This interest would eventually lead to an improved version of the hydrostatic balance for measuring specific gravity. In addition, he developed an understanding of the principle of the pendulum, which he suggested could be used to regulate clocks. It seems that during this period, he also started his careful analysis of the motion of freely-falling bodies. Although there is no evidence that he actually carried out experiments from the top of the Leaning Tower, as is commonly thought, he did devise an ingenious experiment to study objects falling 'almost freely' by rolling them down an incline.

Remarkably, even before finishing his degree, at the age of twenty-six, Galileo was appointed Professor of Mathematics at Pisa. As a professor, he was constantly provoking the older academics and earned himself the nickname of 'The Wrangler'. Not surprisingly, the university did not renew his contract and by the summer 1592, he was out of a job.

By this time, his father had died leaving him with the care of his younger brother Michelangelo. Galileo was thus pressured into finding a new position as soon as possible, in order to provide for his siblings and later his own children. This task he seemed to accept with some resentment. Fortunately, he had established a reputation in the fields of mathematics and mechanics and was able to secure an important appointment at the University of Padua. He carried his reputation with him out of the Duchy of Florence and into the territory governed by that great maritime power, the Serene Republic of Venice, relatively free of the rigid regulations of the Roman Catholic Church.

The eighteen-year period that Galileo spent in Padua he called the happiest of his life. It was in Padua that he made significant discoveries in both pure science and applied science, including testing the strength of materials and improving the telescope. As a lone professor in this field, he taught himself new ways to study the physical world, defining original concepts as he went along.

Furthermore, his personal life changed dramatically in the exciting and wealthy Republic of Venice. He quickly became friendly with a number of Venice's leading citizens. After surviving a bout of financial trouble in early 1593, when the demands of his family and particularly his sister's dowry almost overwhelmed him, he began to prosper. He eventually moved from a small cottage into a larger three-storey house whose grounds included a walled garden where he often entertained students and other guests. In 1599, at the end of his first seven-year term, the university offered to renew his appointment, and Galileo accepted. It is reported that by this time he was one of the highest paid professors on the university faculty.

Although he professed to be a devout Roman Catholic, Galileo fathered three children out of wedlock with a commoner, Marina Gambi, a courtesan from Venice. They had two daughters, Virginia (1600) and Livia (1601), and one son, Vincenzio (1606). Despite their illegitimate status, Galileo recognized his children as his heirs, but he considered the girls unmarriageable, leaving them with only one alternative. Both girls were sent to the convent of San Matteo in Arcetri, outside Florence and remained there for the rest of their lives.

Virginia took the name Maria Celeste upon entering the convent and maintained contact with her father throughout her life by letter. Though none of Galileo's letters survived, Maria Celeste's 120 letters are extant. These heart-felt letters, written between 1623–34, depict Maria Celeste as a woman of brilliance, industry, sensibility and a deep love for her father. She died on 2 April 1634, and is buried with Galileo at the Basilica di Santa Croce in Firenze.

Galileo's Universe

Galileo had learned the traditional Aristotelian view of the universe as a student at the University of Pisa. However, by the time he arrived at Padua, the work of Tycho, Kepler and others were beginning to undermine Aristotle. As we have seen, a new way of approaching science, using mathematics, was developing. Galileo would be a lone explorer, using this new map to chart the unknown heavens. Many of Galileo's academic peers still endorsed old ideas, and strict battle lines were drawn within the university.

Galileo's first priority was to describe mathematically the motion of simple ordinary objects, like that of a ball rolling down an incline. This led to experiments on bodies falling freely in air and soon Galileo developed his particular ability to isolate and focus on the prominent or important aspect of an experiment. For example, closely examining different bodies falling freely from the same height through large distances may indeed show that the bodies do not reach the ground at exactly the same time, a finding seemingly consistent with Aristotle's theory. However, the important point is not that the arrival times are slightly different, but they are very nearly the same.

Galileo thus tried to prove that the slight differences in the arrival times of objects could be attributed to differences in the size, shape and weight of the bodies, which in turn cause unequal air resistance. Ignoring the effects of air, the underlying motion was the same, or so he wished to prove.

Suppositions such as these are counter-intuitive and defied common sense. One experiment to observe the speed at which a penny and a feather would fall in a glass tube from which the air had been evacuated would have shown that the two objects fall at the same rate and land simultaneously at the bottom of the tube. This simple experiment would eventually prove that Galileo was right about free fall, but not until a few years after his death when the vacuum pump was invented. Astronauts dropping objects in the empty atmosphere of the Moon demonstrated that it is air resistance that causes differences in such motion. Since the time of Galileo, the phrase 'in free fall' has been used in physics to refer to objects falling when the only force acting is gravity and when air friction is negligible.

Using a mixture of intuition and mathematical reasoning, Galileo would come to define uniform acceleration as the constant rate of change of velocity over time, which he believed was how freely falling objects moved. The next task was to show that this definition was useful for describing observed motion. It is possible to see how Galileo's simple experiment like that of a ball rolling down an inclined plane, which involved only crude apparatus, turned into a major breakthrough in the study of motion and an important part of the legacy that Galileo left for Isaac Newton.

His commonplace studies of simple motion helped prepare the way for a new kind of physics and later, a new cosmology. Galileo planted the seeds of doubt about the basic assumption of Aristotelian science and by 1609, in addition to his famous free-fall hypothesis, he had determined other new laws of motion. These include his finding that the distance a body falls is proportional to the square of the elapsed time (the law of falling bodies) and that the trajectory of a projectile is a parabola. In addition to his fundamental work on the motion of bodies, he added original work on two major concepts that are fundamental to the final formulation of classical physics as defined by Newton: the subtle idea of inertia and the strange theory of relative motion.

In experiments using a pair of inclined planes facing each other, Galileo observed that a ball would roll down one plane and up the opposite plane to approximately the same height. If smoother planes were used, the ball would roll up the opposite plane even closer to the original height. Galileo reasoned that any difference between initial and final heights was due to the presence of friction. He further postulated that if friction could be entirely eliminated, then the ball would reach exactly the same height.

Galileo further observed that regardless of the angle at which the planes were oriented, the final height was almost always equal to the initial height. If the slope of the opposite incline were reduced, then the ball would roll a further distance in order to reach that original height. He persisted with his reasoning: if the opposite incline was elevated at nearly a zero-degree angle, then the ball would roll indefinitely in an effort to reach the original height. To put this in different words, if the opposing incline was oriented horizontally, then an object in motion would continue to move indefinitely.

This radically modified the generally accepted (including Kepler) notion that a force is required to keep an object in motion. Galileo found the opposite to be true: An object in motion does not come to a rest position because of the absence of a force. Rather, it is the presence of a force – that force being the force of friction or something else – which brings the object to a rest position. In the absence of a force, the object would continue in motion with the same speed and direction – forever!

The natural state of motion – the simplest of physical laws – is therefore that of a body moving with a constant speed in a straight line. Newton would later adopt this principle directly from Galileo as one of his three laws of motion.

The Theory of Relativity
Using his new law of inertia, Galileo was able to explain the natural motion of falling objects on a rotating, or moving, Earth. Galileo proposed that there is no difference between dropping an object from a tower on the surface of the moving Earth and dropping a similar object from the mast of a moving

ship. Because the falling object in both cases shares the horizontal motion of its starting point, the Earth in the first case and the ship in the second case, it appears to drop straight down to an observer who also shares the horizontal motion. The object therefore lands at the base of the tower or the base of the mast, as is expected. But to be exact, the object on the ship has two horizontal motions, that of the ship moving through the sea and the moving Earth. This argument can be taken one step further if you consider that the solar system is not at rest, but is itself moving through intergalactic space.

Galileo ultimately realized that all motion is relative. To be more precise, he helped confirm that it is necessary to select individual frames of reference for each particular experiment involving motion. The French philosopher Pierre Gassendi carried out an experiment in 1640 to show that Galileo was right. This would prove to be the starting point for Albert Einstein, who in 1905 proceeded to demonstrate the bizarre consequences of this principle when the speeds involved in experiments are close to the speed of light. He called this case 'special' relativity.

Galileo's work supported one subtle misconception. To him, a horizontal surface was one that was totally perpendicular to the direction of the centre of the Earth. Thus, his law of inertia would mean that in the absence of any outside force, an object set into motion would ceaselessly continue moving in a circle around the Earth. And this is where Galileo went wrong. Objects set in motion tend to continue moving with a constant speed in a straight line, not in a circle. At a crucial point in Newton's study of planetary motion the English polymath, Robert Hooke, solved the problem. Hooke's discovery was notorious for not being acknowledged by his archenemy in Newton's classic work *The Principia*, as will be seen in the next section.

Galileo and the Telescope

The use of instruments to probe the cosmos is now considered to be fundamental. Early stargazing inspired theorizing which in turn led to new observations – and consequently more theorizing. The early use of the telescope, most famously by Galileo, is a major new chapter in man's ability to explore the universe.

With the exception of primitive eyeglasses, the telescope was the first optical instrument ever constructed; yet its origin is surrounded by controversy. The most likely story places its invention just after 1600, in the shop of an obscure Dutch eyeglass maker named Hans Lippershey. According to an often-repeated story, Lippershey's apprentice was playing with two glass lenses that he had shaped and polished for spectacles. Holding one lens close to one eye and the other at arm's length, the young apprentice inadvertently peered through them in the direction of a distant church steeple. Initially what he saw was a blur, but after adjusting the two lenses and changing the distance between them by moving his arm, the image of the church steeple suddenly snapped into focus. Stunned by the sudden clarity of the steeple, the young man realized that the steeple seemed closer than when he looked at it without the lenses. In excitement, he handed the two lenses to his master, who looked for himself and immediately recognized the significance of the discovery. Lippershey later placed the two lenses at the opposite ends of a long tube, creating a device that he called a 'looker'.

In 1608, Lippershey tried to sell his 'looker' to the Dutch army, but his offer was turned down because of prior claims for the invention. News of the invention spread rapidly. Before the year was out, the French ambassador at The Hague obtained one for King Henry IV and in the next year, lookers were being sold in Paris and some cities in Germany. Soon they appeared in Milan and Venice and shortly after, they were being made in London.

Within a month of Lippershey trying to sell his looker to the Dutch Army, word of the invention reached Venice. It was here that an unidentified stranger tried to sell one to the Venetian Senate, who referred the matter to its scientific adviser, one

Paolo Sarpi, who examined it. When the stranger and his instrument then disappeared, Sarpi went to see the city's most respected instrument maker, none other than Galileo himself, and described the instrument.

Galileo quickly worked out some of the optical principles involved in the looker and set to work, grinding lenses in order to build such an instrument himself. His first telescope made objects appear three times closer than when seen by the naked eye. By the time he had made his third version of the instrument, he was convinced of the tool's groundbreaking ability and considered turning it to the sky.

During a few short weeks in 1609 and 1610 Galileo used his telescope to make several major discoveries, beginning when he turned his telescope to the Moon. Consider the conclusions he was able to draw:

> . . . the surface of the Moon is not smooth, uniform and precisely spherical as a great number of philosophers believe it to be, as with other heavenly bodies. But it is uneven, rough and full of cavities and prominences being not unlike the face of the earth, relieved by chains of mountains and deep valley.

The first telescopic observations of the Moon on record were carried out by the Englishman Thomas Harriot (ca. 1560–1621), on the evening of 26 July 1609, only weeks before Galileo. However, based on his extant correspondence as well as entries in his notebooks, Harriot did not appear to have drawn any particular significance from what he saw, namely, that the surface of the Moon is imperfect. Galileo, on the other hand, was inspired by these simple observations, contrary as they were to the Aristotelian idea of heavenly perfection. But he did not stop there. He supported his observations with several types of evidence, including careful documentation of what he was seeing. For example he worked out a method for determining the height of the mountains of the Moon from the length of their shadows. His value of about 4 miles (6.4

kilometres) for some lunar mountains is not far off modern results.

It is from this reporting that the qualities of investigation which make Galileo stand out from his counterparts become evident. For by insisting on the careful collection of evidence to support his observations, Galileo may be considered the world's first true scientist. In addition, he seemed able to anticipate the criticisms that many would wager against him, that he was creating optical illusions with his telescope and the images in his tube could not be trusted.

Turning from the Moon, the Italian excitedly looked at the stars, only to make another astonishing discovery. Wherever he pointed his telescope in the sky, Galileo saw many, many more stars than appear to the unaided eye. He found that the fuzzy line across the heavens, which we now know as the Milky Way galaxy viewed edge-on, was not a continuous blotchy band of light as it appears to the naked eye, but consisted of thousands of faint stars when viewed through the telescope.

But Galileo's most important discovery was yet to come. He had become obsessed with using his new instrument to view the entirety of the heavens, recording all he saw on each clear night during the winter of 1609. He discovered hundreds of new stars that he knew had never been seen before by another human being. He continued into the New Year and on the evening of 7 January 1610 he made a discovery that he considered to be his most important. Looking through his telescope in the vicinity of the planet Jupiter, he noticed the following:

> That besides the planet there were three 'starlets', small indeed, but very bright. Though I believed them to be among the host of fixed stars, they aroused my curiosity somewhat by appearing to lie in an exact straight line.

When he looked again the following night, the starlets had changed their position with respect to Jupiter but remarkably, were still aligned in a straight line. With each clear evening for

Galileo's notes on observation of Jupiter's
moons, December 1609–January 1610.

weeks after, he observed and recorded the position of these 'starlets' in simple sketches. Within days he concluded that there were four in total and that they were not stars, but satellites of Jupiter.

Of all of Galileo's discoveries, Jupiter's satellites caused the most controversy. However despite this he continued to point his telescope towards other discoveries. By projecting an image of the Sun on a screen he was able to observe sunspots, indicating that the Sun, like the Moon, was not perfect in the Aristotelian sense, and is as flawed and covered with imperfections. Galileo also found that Venus showed all phases, just as the Moon does, indicating that this planet could not always stay between the Earth and the Sun as Ptolemaic astronomers assumed but rather that it must move completely around the Sun as Copernicus and Tycho believed. He also saw that Saturn seemed to carry bulges around the equator. The next generation would confirm this as Saturn's ring, but Galileo's telescope was not yet strong enough to show any detail.

In March 1610, he published his description of his observations of the night sky in a book entitled *The Starry Messenger*. The book was an immediate success, with copies selling as fast as they were printed. This result provoked a great demand for telescopes and great fame for Galileo. Although it contained only twenty-four pages, it astonished and troubled the learned world with its reports that the Moon was not smooth as previously believed, but rather rough and covered with craters; that the Milky Way was made up of faint stars; that Jupiter had its own set of 'moons', or satellites; and that Venus exhibited phases.

In Galileo's opinion, his most important discovery was the previously unidentified four planets that orbit about Jupiter. This discovery contradicted the Aristotelian notion that the Earth is the centre of the universe and therefore the centre of all motion of revolution. Unfolding before his eyes was a miniature solar system with its own centre of revolution.

Galileo was rewarded for his discoveries with life tenure at Padua and a doubling of his salary for his success with the telescope. On 14 April 1611, about a year after the publication of his sensational *Starry Messenger*, he was feted at a banquet held in his honour near Rome. After showing the guests how he had made his discoveries, an unidentified Greek poet-theologian proposed a name for the instrument, one borrowed from ancient Greece. His suggestion was quickly accepted and that same night the host, Federico Cesi, officially christened Galileo's instrument 'the telescope'.

Kepler and Galileo, Four Hundred Years Ago

UNESCO's celebration of the joint contribution to astronomy of Kepler and Galileo begs us to consider whether these two Renaissance figures ever met. The answer is no. They were separated both by the Alps and that more formidable of barriers, religious intolerance. However, a study of their well-documented lives reveals some remarkable aspects of their differing personalities.

Though Kepler and Galileo lived during the same era, their lives were completely different. The Italian was born in 1564, seven years before Kepler and outlived the German by twelve years, dying in 1642. As we have seen, Kepler was raised in near poverty and, as an adult, was driven from city to city by religious wars. His work was unknown by his contemporaries and even today he is not fully appreciated for the contribution he made to the scientific revolution.

Galileo, by contrast, came from an aristocratic family and spent his whole life in Catholic Italy: Florence, Pisa, Padua, Rome and finally under house arrest at Arcetri, just outside Florence. Another important difference was in the way the two reported their work. Kepler wrote complicated books in scholarly Latin that demanded expert knowledge from his readers. The Italian, on the other hand, wrote his essays and books in the vernacular, making them accessible to most readers. Their personalities could likewise not be more different. Kepler was a kind, sensitive, almost mystical character who never felt the need for competition, selfishness or self-promotion. Galileo was opposite in all of these respects.

Such utter disparity between these two giants of Renaissance science is most pronounced when reviewing their correspondence, which began in 1597 when Kepler sent Galileo a copy of *Mysterium Cosmographicum*. Their interaction ended during the 1609, the year that Kepler published his *Astronomia Nova* and Galileo reported on his telescopic investigations.

In 1597, Kepler had sent copies of his *Mysterium* to astronomers all over Europe, including Tycho and Galileo, who was at that time establishing himself at Padua. After waiting months for an opinion of his book from Italy, Kepler received a reply on 4 August 1597:

Your book, most highly learned gentleman, has just reached me not days ago but only a few hours ago and I should think myself indeed ungrateful if I should not express to you my thanks by this letter. I thank you especially for having deemed me worthy of such a proof of your friendship . . . So far I've read only the introduction, but have learned from it in some measure your intentions and congratulate myself on the good fortune of having found such a man as a companion in the exploration of the truth. For it is deplorable that of that there are so few who seek the truth and do not pursue a wrong method of philosophising. But this is not the place to mourn about the misery of our century but to rejoice to few about such beautiful ideas proving the truth. So I add only this promise that I will read your book in peace, for I'm certain that I will find the most beautiful things in it . . . I would certainly dare to approach the public with my ways of thinking if there were more people in your mind. As this is not the case, I shall refrain from doing so. The lack of time and the ardent wish to read your book make it necessary to close, assuring you my sympathy. I shall always be at your service. Farewell, and do not neglect to give me good news of yourself.

Yours in sincere friendship,
Galilaeus, Mathematician at the Academy of Padua.

The letter offers little suggestion of Galileo's later scandalous treatment of Kepler. Unfortunately, the naive and trusting German had no idea he was being conned by one of the great self-promoters in history. Galileo, it transpired, had no intention of being 'always at your service'.

Nevertheless, the unassuming Kepler attempted to continue his correspondence with Galileo and the following month wrote:

> I received your letter of August 4th on the first day of September. It was a double pleasure for me. First because of [becoming] friends with you, the Italian, and second because of the agreement which we find ourselves concerning Copernican cosmography. As you invite me kindly at the end of your letter to enter into correspondence with you, and I myself feel greatly tempted to do so, I will not let pass the occasion of sending you a letter with the present young nobleman. For I am sure, if your time has allowed it, you had meanwhile obtained a closer knowledge of my book. And if so a great desire has taken hold of me, to learn your judgment. For this is my way, to urge all those to whom I've written to express their candid opinion. Believe me, the sharpest criticism of one single understanding man means much more to me than the thoughtless applause of the great masses . . .

Kepler continues the letter, encouraging Galileo to join the public debate on the validity of the Copernican scheme. In a somewhat prescient phrase he makes reference to problems with the Roman Church which were to famously torment the Italian in later years:

> Be of good cheer, Galileo, and appear in public. If I am not mistaken there are only a few amongst the distinguished mathematicians of Europe who dissociate themselves from us. So great is the power of truth. If Italy seems less suitable for your publication and if you have to expect difficulties there, perhaps Germany will offer us more freedom. But enough of

this. Please let me know privately if you do not want to do so publicly, what you have discovered in favour of Copernicus. Farewell and answer me with a very long letter.

Galileo never answered with 'a very long letter' or indeed with any other length of letter. He ignored Kepler completely, though it was later reported that the Italian incorporated some of the new ideas from *Mysterium* in lectures at Padua as if they were his own. Many years later, he ridiculed Kepler's style of writing and mocked his notion that tides were caused by the attraction of the Moon – a theory later shown to be correct. Galileo was clearly never one to accept another's ideas as superior to his own.

Over a decade after this exchange of letters, when Kepler was Imperial Mathematician in Rudolf's court in Prague, he wrote in support of Galileo's 'remarkable telescopic observations of the heavens' (*The Starry Messenger*). Yet, this did not move the ambitious Italian ever to answer Kepler's requests for one of his telescopes. These Galileo mostly distributed to his many aristocratic patrons; naming the newly discovered Moons of Jupiter after his patrons, the Medicean princes, as well.

The ambitious, calculating scientist was too busy with his own self-promotion to exchange any new scientific ideas or instruments which someone whom he knew to be a scientific genius and who one day might compete with him for a position of patronage from his own benefactors. Generosity was a completely foreign concept to Galileo – Kepler would certainly be the last person to receive any of Galileo's unpublished work or a telescope. It is fair to say that the Italian was intimidated by Kepler's brilliance, and once admitted that there was only one man who knew more than he did about the principles of the telescope.

At this point, even after Kepler's enthusiastic support for *Sidereus Nuncio*, the personal contact between the two ceased. In subsequent months, Kepler wrote several more letters that

Galileo left unanswered. Furthermore, in his work Galileo rarely mentions Kepler's name. When he does it is only with the intent to refute certain ideas, as we saw in the earlier example of the effects of the Moon on the tides. Kepler's three laws, his 'physical' astronomy and his work on optics were all ignored by Galileo, who, until the end of his life, defended the idea that circular orbits and epicycles were the only conceivable form of heavenly motion. He thus ignored the whole basis for the development of modern astronomy: Kepler's elliptical orbits of the planets and the concept that the Sun emits a physical force which in turn causes the planets to move in their orbits.

Albert Einstein was one of only a few important scientists who ever criticized Galileo for his treatment of Kepler, remarking near the end of his life that:

> It has always hurt me to think that Galileo did not acknowledge the work of Kepler . . . alas, that is vanity. You find it in so many scientists.

Galileo's Final Battle

Galileo's problems with the Church had little to do with developing any understanding of the universe. In 1612, just after cutting himself off from Kepler, opposition to the Sun-centred theory of the universe arose. Though Galileo did support the theory, the version he supported was a primitive Copernican form that didn't include any of the new ideas from Kepler's *Astronomia*. In 1614, from the pulpit of Santa Maria Novella, Father Tommaso Caccini (1574–1648) denounced Galileo's opinions on the motion of the Earth, judging them dangerous and close to heresy. He had to be careful.

This started a most distressing chapter of the Italian's life, for rather than accepting the censure, Galileo went to Rome to defend himself against the accusations levied against him. Despite his best efforts, in 1616 Cardinal Roberto Bellarmine handed Galileo an admonition enjoining him neither to advocate nor teach Copernican astronomy. During 1621 and 1622

Galileo wrote his first book, *The Assayer* (Il Saggiatore), which was approved and published in 1623. In 1630, he returned to Rome to apply for a licence to print his most important work, the *Dialogue Concerning the Two Chief World Systems*, which was then published in Florence in 1632. In October of that year, at the age of sixty-eight, he was ordered to appear before the infamous Holy Office in Rome.

Popular mythology sees Galileo as a 'martyr of science', a man persecuted by the Church for his scientific beliefs and, in particular, for teaching that the Earth moves around the Sun. Yet historical research shows that this is not true. Galileo, we must not forget, was a provocative and proud man, with an established track record of rows and lawsuits. But it was not until 1616 that the Church expressed any opinions about science, and only in 1632 that the Inquisition questioned Galileo. This was eighty-nine years after Copernicus' ideas had been in free circulation across both Catholic and Protestant Europe.

That Galileo was an astronomer of the greatest importance is beyond any doubt. After 1610, bishops and cardinals applauded Galileo's telescopic discoveries, and Galileo travelled to both Florence and to Rome as a celebrity. So what happened? In 1616 he got into trouble for beginning to teach Copernicus as the sole acceptable truth. The heresy for which he was condemned in 1632 was therefore one of academic disobedience, not religious. Until the day he died in 1642, no one doubted the sincerity and orthodoxy of his Christian beliefs. So if we insist on seeing Galileo as a martyr of science, we must not forget that he also courted controversy with the Church. 'The Wrangler' could simply not walk away from an argument.

Nevertheless, following a papal trial in which he was found vehemently suspect of heresy, Galileo was placed under house arrest and his movements restricted by the Pope. From 1634 onward he stayed at his country house at Arcetri, outside of Florence. He went completely blind in 1638 and after suffering from a painful hernia and insomnia, was permitted to travel to

Florence for medical advice. He continued to receive visitors until 1642, when, after suffering fever and heart palpitations, he died on 8 January 1642.

The Grand Duke of Tuscany, Ferdinando II, wished to bury him in the main body of the Basilica of Santa Croce, next to the tombs of his father and other ancestors, and to erect a marble mausoleum in his honour. These plans were scrapped, however, after Pope Urban VIII and his nephew, Cardinal Francesco Barberini, protested. He was instead buried in a small room next to the novices' chapel at the end of a corridor from the southern transept of the basilica to the sacristy. He was reburied in the main body of the basilica in 1737 after a monument to Michelangelo had been erected there just opposite the tomb.

The Inquisition's ban on reprinting Galileo's works was lifted in 1718 when permission was granted to publish an edition of his works (excluding the condemned *Dialogue* in Florence). In 1741 Pope Benedict XIV authorized the publication of an edition of Galileo's complete scientific works, which included a mildly censored version of the *Dialogue*. In 1758 prohibition against works advocating heliocentrism was removed from the Index of Prohibited Books, although the specific ban on uncensored versions of the *Dialogue* and Copernicus' *De Revolutionibus* remained. The Church's last traces of official opposition to heliocentrism disappeared only in 1835, when these works were finally dropped from the Index.

Within a few months of his election to the papacy in 1939, in his first speech to the Pontifical Academy of Sciences, Pope Pius XII described Galileo as being among the 'most audacious heroes of research . . . not afraid of the stumbling blocks and the risks on the way, nor fearful of the funereal monuments'. Pius XII was very careful not to close any doors (to science) prematurely. He was energetic on this point and regretted what had occurred in the case of Galileo. On 15 February 1990, in a speech delivered at the Sapienza University of Rome, Cardinal Ratzinger spoke of the Galileo affair as forming what he called 'a symptomatic case that permits us to see how deep the self-doubt of the modern age, of science and technology goes today.' On 31 October 1992, Pope John Paul II expressed regret for how the Galileo affair was handled, and officially conceded that the results, of a study conducted by the Pontifical Council for Culture, show the Earth is indeed not stationary.

8

ISAAC NEWTON AND THE SCIENTIFIC REVOLUTION

Isaac Newton was born on Christmas Day in 1642, the year that Galileo died. His family owned a farm in the small English village of Woolsthorpe in Lincolnshire, about 100 miles (161 kilometres) north of London. Born tiny and weak, Newton was not expected to survive his first day of life. His father died before his birth and Newton grew into a quiet farm boy who experienced a troublesome childhood. After two years as a widow, his mother married a second time to the well-to-do minister, Barnabas Smith, moving to a nearby village to raise the minister's son and two daughters. Young Isaac was made to live with his grandmother for nine years, until the death of the minister in 1653. Biographers attribute Isaac's traumatic separation from his mother as the cause of his pronounced psychotic tendencies in later life.

As a young boy Newton enjoyed the tangible act of building mechanical devices such as clocks and windmills, much as Galileo had done as a youth. At the same time, he demonstrated a liking for the abstraction of mathematics, a rare combination. A total lack of interest in farming led him to accept financial help from an uncle so that he could attend Trinity College in Cambridge University in 1661. There he enrolled for a degree in mathematics.

By the time Newton had arrived in Cambridge, the movement known as the 'new philosophy' was underway, and many of the works basic to modern science had already appeared. Copernicus and Kepler had elaborated the heliocentric system of the universe and Galileo had proposed the foundations of a new mechanics built on the principle of inertia. In addition, philosophers such as René Descartes in France had begun to formulate a new conception of nature as intricate, impersonal and inert.

But at Cambridge, none of these advancements had any bearing on Newton's education. The university instead continued to be a stronghold of outmoded Aristotelianism, which rested on a geocentric view of the universe and dealt with nature in qualitative rather than quantitative terms. With aspirations to be a successful student, and despite his awareness of the new ideas that were taking shape around him, the young provincial farm boy immersed himself in Aristotle's writings. Ominously, he wrote in his personal notebook dating from this period that, 'Plato is my friend, Aristotle is my friend, but my best friend is truth.'

When Newton received his BA in April 1665, it was without much recognition. What is remarkable about the time that Newton spent at university is his independent acquisition of those books containing the 'new' philosophies and mathematics of his time. No one knew of his independent intellectual pursuits, for he confined his study and progress with these ideas to the writing he did in his personal notebooks.

Shortly after receiving his degree, Newton was made aware of a severe epidemic sweeping across England. Though the bubonic plague had been around for centuries, the one that struck in 1665, the Great Plague, proved to be a particularly severe manifestation. London was the worst affected, but all locations where people gathered were advised to close, and so the colleges at Cambridge shut their doors for two years.

During the years when the university remained closed, Newton went home to Woolsthorpe and for two years, he consolidated and expanded his independent studies. In mathematics, he discovered the binomial theorem and also differential calculus, which he called 'the theory of fluxions'. In optics, he worked out a theory of colours and in mechanics formulated a clear concept of his first two laws of motion. A further claim, made sixty years later, that he had also conceived of his law of universal gravitation during this period when he watched an apple fall to the ground, was later found to be a deceit (as we will see). He did not arrive at the correct notion about gravity until 1679.

However in an important advancement toward understanding planetary motion, he did derive the equation for the acceleration of a body moving in a circle during this period. However, one essential point confused him, namely the direction of the force producing the acceleration, which he conceptualized as an outward force. This kept him from understanding how the Sun attracted the planets.

In 1667 after the university reopened, Newton was immediately elected to a fellowship in Trinity College. Two years later, the Lucasian Professor of Mathematics, Isaac Barrow, resigned his position as chair to devote himself to divinity studies. In a remarkable gesture, he recommended Newton as his successor. This full professorship, at the young age of twenty-seven years old, exempted Newton from his tutoring obligations, requiring him only to deliver an annual course of lectures. He chose the work he had done in optics as the initial topic for these lectures.

The Theory of Colours

The core of Newton's contribution to the field of optics was centred on colours. An ancient theory extending back at least to Aristotle held that a certain class of colour phenomena – that appearing in rainbows, for example – arises from the modification of light. According to this model, light appears white in its pristine, or unmodified, form. But through a series of experiments performed in 1665 and 1666, Newton produced a spectrum of colours by projecting a narrow light beam onto the wall of a darkened chamber.

His findings from this experiment led him to deny the concept of modification and caused him to refute the idea that light is simple and homogeneous in its unmodified form. He postulated instead that light is complex and heterogeneous and that colours arise when this heterogeneous mixture is broken down into its simple components. He further contended that the mixture's component rays excite the sensations of individual colours when they strike the retina of the eye.

Newton also concluded that different rays refract, or bend, at distinct angles in glass, hence the spectrum produced by a prism. Because he believed that chromatic aberration (different refractions cause different colours) could not be eliminated from lenses, he also criticized the simple refracting telescope. He therefore proposed a new design; a reflecting telescope where the light captured by the tube is reflected to the focus point. He proceeded to construct the first reflecting telescope, later demonstrating it to the Royal Society. Yet, in spite of his efforts, there is no evidence that his theory of colours, fully described in his inaugural fellowship lectures at the university, made any impression. Fortunately, a new and important forum for the spread of scientific ideas had recently been established.

The Royal Society

The Royal Society of London, which is today the oldest national scientific society in the world and the leading national organization for the promotion of scientific research in Britain, was founded on 28 November 1660. On that day, twelve men met after a lecture given by Christopher Wren, who was then the professor of astronomy at Gresham College, London. These men resolved to set up a College for the promotion of 'Physico–Mathematical–Experimental Learning'. These ambitions were put into effect over the next few years, particularly through a charter of incorporation granted by King Charles II in 1662.

Newton chose to transmit his work on optics and colours through the Royal Society. His name was otherwise unknown to these men of science, even after his appointment as Lucasian professor. But this changed in 1671, when the Society heard about Newton's reflecting telescope and asked to see it. Pleased by the Society's enthusiastic reception of the telescope and by his election to the Society, Newton volunteered a detailed paper on light and colours in 1672.

On the whole, the paper was well received by the Society, though some dissent was registered. Robert Hooke, Curator and Senior Experimenter of the Royal Society, who considered himself a master in optics, wrote a condescending critique of Newton's new theory. The rage provoked in Newton by Hooke's critique was astonishing. It was the first example in a series of incidents of Newton's inability to accept criticism in a rational manner, especially if the criticism came from Hooke. Less than a year after submitting the paper and unsettled by the continuous disagreement, he began to cut his ties not just from the Royal Society, but also from the entire scientific community. Newton soon withdrew into virtual isolation.

Hooke's Big Idea

Newton's pioneering work on light and colour is not a major concern in this book, but his introduction to Hooke and others through the Royal Society certainly is. Seven years after Newton's election, Hooke – who was now Secretary of the Royal Society – made another overture to Newton. At Cambridge, Newton had been developing the idea of attraction and repulsion with regard to terrestrial phenomena. Late in 1679, seeking to renew their correspondence Hooke suggested a different application of the attraction/repulsion phenomenon in a letter to Newton. Hooke mentioned Newton's new analysis of planetary motion, asking Newton if he had ever considered *the continuous diversion of the straight tangential motion of a planet by a central attractive force*. By this, Hooke was referring to a situation in which a planet moving in a straight line near the Sun would be bent into a curved path by the Sun's pull.

Newton bluntly refused to correspond about this problem, but went on to mention a different experiment about the rotation of the Earth – he didn't want to suggest to Hooke that this new idea about planetary motion might be an important one. Hooke's letter ultimately caused Newton to think about orbital dynamics in 1679–80, well before he arrived at the concept of universal gravitation.

It is important here to examine the importance of Hooke's idea in the chronology of Newton's great discovery of universal gravitation. This takes us back to Newton's claim that his discoveries were made in the 1660s – and unveils new information about his ubiquitous apple!

Revelations at Harvard

Much has been written about Sir Isaac Newton. Of these volumes, any thorough examination of the work of the great English polymath that contains more than a hint of scandal risks being considered newsworthy. It is equally scandalous to suggest that Newton's work in the area of gravitation and planetary motion may be in any way slightly overrated or even partly plagiarized. Yet it seems reasonable to suggest that Sir Isaac appropriated fundamental ideas from Kepler, Galileo and, particularly, Robert Hooke with little acknowledgement of any of these men.

Physicists and teachers of science have always been interested in Newton's work because it presents a formidable example of the wonderful order and structure of the subject of physics. In addition, Newton's well-known apple story epitomizes the excitement of discovery in science – the *Eureka* moment. Yet it is difficult to accept certain questionable elements about the discovery of gravitation that Newton and his acolytes have managed to promulgate. First, he claimed to have worked out the entire solution of gravity as a student in the 1660s, and secondly, that the basic inspiration for his idea came from a falling apple.

According to these accounts, this all took place on his family farm in Lincolnshire where he went to escape from plague-infested Cambridge University in 1665–66. He supposedly related this information to William Stukeley, an English antiquary known for his work on the archaeological investigation of Stonehenge and Avebury, and one of Newton's earliest biographers. Newton met Stukeley some sixty years after the plague in 1726, a year before he died. At that meeting, he claimed that the solution for the force of gravity came to him during that early period on the farm, more precisely, at a moment while sitting in his garden when he saw an apple fall from a tree to the ground. As Stukeley quoted Newton:

> . . . the notion of gravitation came into my mind when I was occasion'd by the fall of an apple, as I sat in a contemplative mood. Why should the apple always descend perpendicularly to the ground, I thought to myself. Why should it not go sideways or upwards, but constantly to the Earth's centre?
>
> I began to think of gravity extending to the orb of the moon and . . . from Kepler's rule [the law of periods] . . . I deduced that the forces which keep the planets in her orbs must be reciprocally as the square of the distances from the centre on which they revolve: and thereby I compared the force requisite to keep the moon in her orb with the force of gravity at the surface of the earth and found them to answer pretty nearly [i.e. the same]. All this was in the two plague years of 1655 and 1666, for in those days [age twenty-one and twenty-two] I was in the prime of my age for invention, and minded mathematics and philosophy more than any other time since.

Many people with training in science history question the perpetuation of what is known as the apple story. 'Newton's Discovery of Gravitation' (1981), an article published in *Scientific American* by I.B. Cohen, a distinguished Harvard historian of science, states that Newton fabricated much of the information on his discovery of gravitation and then predated the facts to guarantee priority for the discovery.

Cohen made two surprising claims in his *Scientific American* article concerning Newton's apple story. Firstly, that Newton did not make the discovery of gravitation until he was forty-two years old, some twenty years later than is commonly believed; and secondly, that he was only able to make the discovery after his arch-rival Robert Hooke clarified the essential concept of curved motion in his letter of 1679.

Myriad theories about Newton's apple story abound. For many years it was thought to be an invention by historians, with French writer Voltaire often cited as the chief suspect. Another version by the late J.B.S. Haldane, an outstanding British biological scientist who acted as editor of the *Daily Worker* during his final communist days, suggests that the story of the apple was capitalist propaganda.

As we have seen, Newton's own account of the story was only revealed at the time of the Second World War, when the unpublished manuscript biography of Newton by Stukeley was discovered. The great scientist was an old man in his seventies, when Stukeley interviewed him in his garden in Kensington. During the interview Newton confirmed that he had indeed made his great discovery, watching an apple fall to the ground on his farm at Woolesthorpe.

There are many who believe that such a beautiful story should not be destroyed by any quest for historic accuracy. The tale is iconic, a popular image of a great discovery and has come to epitomize genius. In this respect the important thing is that the basic idea is retained, so they say. However, Cohen – arguably the dean of Newtonian scholarship in the twentieth century – is quite certain that gravitation was not discovered as Newton dictated it. Furthermore he claims that by accepting the oversimplified apple story, Hooke's great contribution is overlooked.

There is no doubt that Hooke antagonized Newton. As a result the Cambridge professor was never going to credit Hooke with helping him to develop the theory of gravitation or with discovering the inverse square law. In the latter case,

Newton was arguably not being unfair in ignoring Hooke. The assertion that the gravitational force between two bodies decreases as the reciprocal squared of the distance between the two bodies, could after all, have been easily extrapolated from Kepler. His discovery that any quantity that spreads out uniformly from a centre point, like a point source of light, will decrease in intensity as the square of the distance. This is a result of the fact that the surface area of a sphere is related to the square of its radius.

So Newton, Hooke or any scientist of that era would have benefitted from the insight that gravitation law must comprise the inverse square form. Once it had been discovered that the centripetal acceleration (or 'centrifugal acceleration' as it was mistakenly called in Newton's day) varies as the velocity V-squared divided by the radius R, then Kepler's third law for circular cases could be applied to obtain the inverse square law without any difficulties. Newton would have attempted this particular calculation for the planets as a young man.

Newton's notebooks show that he had derived the V-squared over R relation for centripetal acceleration back in the 1660s, though his proof wasn't published until a decade later by the Dutchman Christian Huygens. Newton did find that formula in the sixties and then used Kepler's third law to infer the inverse square law for gravitation. So he thought that once Huygens had published his *Horologium oscillatorium* in the early 1670s, anyone with a little algebra could find the inverse square law for circular orbits. Consequently, he did not consider it an important discovery and was reluctant to give Hooke any credit.

However, the correct analysis of motion in a curved orbit, the subject of Hooke's 1679 letter, was a different story. Before then no one knew how to analyse such motion. Newton was thinking in terms of a very intuitive concept of centrifugal force. As we now know, this approach is misleading because it concentrates on the experience of the moving body trying to force itself outwards rather than inwards towards the source

Basic physics of centripetal motion incorrectly represented.

of the curvature at the centre. It is essential here to distinguish between a *centrifugal* force, meaning *away* from the centre, and a *centripetal* force, meaning *towards* the centre. The difference is critical to this discussion.

Now to return to 1679, this time in more detail. As shown in 1679 Hooke had just been elected Secretary of the Royal Society and he wrote a letter to Newton saying that he hoped the two men could have a private philosophical correspondence. He knew that Newton was annoyed because of an earlier exchange that the two men had had in 1672, when Newton published his famous theory of light and colours. Newton had replied in a very friendly manner to Hooke – an interesting starting point.

In the course of that correspondence, Hooke asked what Newton thought of his idea of analysing curved motion like the Moon and the planets as the combination of two independent effects: an inertial motion along the tangent to the curve in a straight line and secondly, a force directed toward the centre, making the object curve in its path. Newton replied that this was something he had never heard of before and so he really didn't have any opinion about it.

Newton then suggested a different problem, one of his own, which he described incorrectly. Hooke immediately corrected Newton's error and then returned to his previous question of the new form of analysing curved motion. There is a strong possibility that Newton's new problem, and possibly even his error, was a deliberate attempt to divert Hooke from further discussion on curved motion. This new idea may well have triggered in Newton's mind the entire solution to orbital motion, revealing that physical attraction between the central body and the orbiting body occurs along the radius, which is the only force present.

Hooke replied that by accepting his new way of analysing curvatures, the inverse square law could be obtained. But Newton did not reply to that directive. Hooke then said that Newton should really use his talents to find the law that governs the motion of the planets because of its important implications for navigation. But Newton refused to reply. Here, even in spite of the antipathy they had for each other at that time, Hooke recognized that Newton was the better mathematician and had the skills to analyse that which he could not.

Hooke knew that he belonged amongst the world's greatest inventors, scientists and microscopists – but he also knew that he wasn't one of the world's greatest mathematicians. He could not solve the problem. So he suggested that Newton try.

It seems possible that Newton did the necessary mathematics immediately, applying the new idea about the tangential motion in combination with the central attractive force. Unfortunately, there is no documentation in support of this because that would have meant that Newton would have had to announce a great discovery just after receiving a letter from Hooke. It would have been too obvious. He instead replied that Hooke should perhaps pursue such important work himself. In any event, it seems certain that Newton solved the problem then and there, as suggested on the following grounds.

One of the most famous meetings in the history of science took place in the summer of 1684 when the young astronomer Edmund Halley paid a visit to Newton. During this meeting Halley asked Newton what path a planet would follow if it were attracted to the Sun by a force proportional to the reciprocal of the squared distance. Recall that the idea that the planets were attracted to the Sun by an inverse-square force law had by then occurred to several people, including the architect Christopher Wren, the scientist Robert Hooke and to Newton himself.

Wren, Hooke and Halley had discussed the problem at a coffee house following a meeting of the Royal Society in January of 1684. During the discussion Wren had offered a cash prize to whomever could provide a derivation of the shape of planetary orbits under the assumption of an inverse-square central force of attraction toward the stationary Sun. Hooke had claimed to have a proof that the paths were ellipses, but never provided it. Against this background, Halley paid Newton a visit.

Abraham de Moivre was a French mathematician and pioneer in the development of analytic trigonometry and in the theory of probability. He spent much of his life in London and was a professional friend of Newton. Sometime in the early eighteenth century, Newton told the Frenchman about his meeting with Halley. According to de Moivre:

Dr Halley came to visit Newton at Cambridge. After they had been some time together, the Dr asked him what he thought the curve would be that would be described by the planets supposing the force of attraction towards the sun to be reciprocal to the square of their distance from it. Sir Isaac replied immediately that it would be an ellipse. The Doctor, struck with joy and amazement, asked him how he knew it. Why, saith he, I have calculated it. Whereupon Dr Halley asked him for his calculation without any f[u]rther delay. Sir Isaac looked among his papers but could not find it, but he promised him to renew it and then to send it him. . .

As is well known among science historians, Halley's question prompted Newton to formulate ideas about mechanics and the universe. At the infamous meeting in Cambridge, Halley extracted Newton's promise to send the mathematical demonstration that an inverse square force produces elliptical orbits. Three months later he received a short tract entitled *De Motu* (On Motion) with the promise that Newton was already at work improving and expanding the treatise. In two and a half years, *De Motu* grew into *Philosophiae Naturalis Principia Mathematica* – Newton's masterpiece and one of the fundamental works in the entirety of modern science.

Significantly, *De Motu* did not state the law of universal gravitation. What is more, even though it was a treatise on planetary dynamics, it did not contain any of the three Newtonian laws of motion. Only when revising *De Motu* did Newton embrace the principle of inertia (the first law) and arrive at the second law of motion, the force law, which proved to be a precise quantitative statement of the action of the forces between bodies. By quantifying the concept of force, the second law demonstrated the exact quantitative mechanics that has been the paradigm of natural science ever since. Newton proved not only that the inverse square law is the cause of the elliptical orbits of the planets, but also he eventually solved nearly all the problems relating to celestial and terrestrial motion.

The answer Newton prepared for Halley grew, becoming progressively more comprehensive until, in a remarkably short time (about 18 months), he had composed the three-volume work entitled *The Mathematical Principles of Natural Philosophy*. Usually referred to by its Latin title *Philosophiae Naturalis Principia Mathematica*, or simply, *Principia*, this work comprises the foundation of modern physics and represents arguably the greatest single advance in human understanding ever achieved in the history of science.

A number of documents exist in which Newton indicates that his discovery was predicated on the contact that he had with Hooke. Of these, some describe Newton's attempt to write the chronology of events that led to his great discovery in which Hooke earns no credit. In some cases, he specifically mentions certain letters without actually naming Hooke. By his own admission, then, this period of time was influential in his life.

Newton indicates that he learned something new at the time of the correspondence with Hooke in his position as the Secretary of the Royal Society. All of this correspondence is preserved, and as Newton says in his first reply to Hooke, the analysis of curved motion combining inertial tangential motion and attraction towards the centre is something new.

Realizing more than anyone the importance of what Newton was creating, Halley arranged for the publication of *Principia*, going so far as to contribute his personal funds to help meet the printing costs. The result was the *Principia* published in 1687. In the 'Ode to Newton' which prefaced *Principia*, Halley made the following comment: 'Nearer the Gods no mortal may approach.'

One immediate application of Newton's new theory was the description of the elliptical orbit of the comet of 1682, which Halley had been studying. Using Newton's results, Halley discovered that this comet had the same orbit as that of 1607. He then examined reports of a comet seen in 1531, and found the same orbit again. He announced that these seemingly

different comets were, in fact, one and the same object returning to the Earth's vicinity about every 76 years. Based on this
orbital information, Halley predicted that the comet would be
visible again in 1758.

Halley died in 1742, outliving Newton by fifteen years,
but the world's astronomers did not forget his prediction.
Disappointment mounted throughout the appointed year,
when almost a whole year passed without a sighting. Finally,
on Christmas night, Johann Palitzsch, a Dresden amateur
astronomer became excited by an object he saw in the night
sky. It was indeed Halley's Comet, as it then became known,
returning to the inner solar system from the far reaches of its
elliptical orbit.

Halley was a man of many talents. He excelled in mathematics, astronomy and physics, and is considered to be the
founder of geophysics for his early work on winds and tides.
He was the first to establish an astronomical observatory in the
southern hemisphere on the island of St Helena from where
he catalogued the stars in the southern skies. He was active in
the Royal Society for which he was Secretary for many years
and later undertook a variety of assignments for the crown,
including voyages to the south Atlantic to investigate magnetic
compass bearings. These activities assured him a position as
professor of astronomy at Oxford, a position once held by
his friend Christopher Wren, who first set him on the road to
Cambridge to ask for Newton's help. For the last twenty years
of his life he served as Astronomer Royal at the Greenwich
Observatory in London.

The *Principia*

The *Principia* is an extraordinary document. Its three main
sections contain a wealth of mathematical and physical discoveries. But overshadowing everything else is the theory of
universal gravitation, together with the proofs and arguments
that support the theory. Newton uses an unusual form of argument patterned by Euclid, a style of mathematical reasoning

mostly unfamiliar today. Many of the steps used in his proofs are therefore more understandable when restated in modern terms.

The central idea of universal gravitation can be simplified: every object in the universe attracts every other object. The amount of attraction depends in a simple way on the masses of the object (how much matter contained therein), and the distance between the objects.

This represents a great new synthesis, combining terrestrial laws on force and motion with laws of motion derived from astronomy, as in Kepler's laws. Gravitation is a universal force and applies to the Earth (as well as to apples), the Sun, planets and to all other bodies moving in the solar system – even comets. Newton thus united heaven and Earth in a grand system dominated by the law of universal gravitation. The reaction of those who could understand its full implication are perhaps best captured with the words of English poet Alexander Pope who wrote on Newton's epitaph:

> Nature and Nature's laws lay hid in night:
> God said, Let Newton be! and all was light.

The *Principia*, written in Latin, is filled with long geometrical arguments and is very difficult to read and understand. Fortunately, several gifted contemporary writers wrote clearer versions, allowing a wide circle of readers to be exposed to Newton's arguments and conclusions. The French philosopher and reformer, Voltaire, published one of the most popular of these books in 1736.

In the derivation of his universal gravitation law, Newton was able to show that the natural straight-line motion of a planet was forced into a curve by the influence of the Sun. Newton's analysis proved that Kepler's laws could also be derived from universal gravitation. Thus, though the new theory had wider applicability, Kepler's laws were still correct. The equations indicated that a body moving under a central force would move according to Kepler's law of areas and that motion in an elliptical path would occur only when the central force was an inverse square force proportional to the reciprocal of the distance squared.

Finally, Newton showed with mathematical proofs that only an inverse square law force exerted by the Sun would result in the observed elliptical orbits described by Kepler. Recall that this was also the answer that Newton gave to Halley in August 1684, the answer that set in motion the remarkable journey to the *Principia*. Everything was now consistent, with one general law of universal gravitation applicable to all bodies moving in the solar system.

The following is a summary of Newton's three laws of motion, an exact quantitative description of the motion of visible bodies, as stated in the *Principia*:

- That a body remains in its state of rest (or in uniform motion in a straight line) unless it is compelled to change that state by an external force.
- That the change of motion (the change of velocity times the mass of the body) is proportional to the force impressed.
- That to every action there is an equal and opposite reaction.

The analysis of circular motion in terms of these laws yielded a formula of the quantitative measure, in terms of a body's velocity and mass, of the centripetal force necessary to divert a body from its rectilinear path into a given circle. This was based on the key idea that originated with Hooke and allowed Newton to understand how his new concept of a 'force' could be employed to explain the curved motion of the planets. When Newton substituted this formula into Kepler's third law, he found that the centripetal force holding the planets in their given orbits about the Sun must decrease with the square of the planets' distances from the Sun. Thus, Newton had accomplished what Halley had asked, as well as its converse. In so doing he also indicated the complete generality of Kepler's laws.

After *Principia*

After *Principia*, Newton appears to have had a nervous breakdown. He did recover, but from that time until his death thirty-five years later, he made no major scientific discoveries and instead dabbled in writing on alchemy, theology and other aspects of the occult. In 1697 he was appointed to be Master of the Royal Mint in London, partly because of his great knowledge of the chemistry of metals.

During his final years, Newton received many honours and famously represented Cambridge University in Parliament. In 1705, Queen Mary knighted him and he served as President of the Royal Society from 1703 until his death in 1727. Newton was buried with much fanfare in Westminster Abbey and Alexander Pope gave his eulogy. There his statue rests, sitting in repose, leaning on a pile of his own books and accompanied by angels like a god.

The death of Newton brought to an end a glorious period of discovery in physical science that lasted less than two hundred years and yet formulated the methods and models on which most of subsequent science is based. Historical sceptics may claim that the scientific revolution never happened, but many others contend that it did, including British historian Herbert Butterfield, who wrote in 1948:

> The Scientific Revolution overturned the authority not only of the Middle Ages but of the ancient world – it ended not only in the eclipse of scholastic philosophy but in the destruction of Aristotelian physics. The Scientific Revolution outshines everything since the rise of Christianity and reduces the Renaissance and Reformation to the rank of mere episodes, mere internal displacements within the system of medieval Christianity. It heralded a new dawn of possibilities and a new world governed by laws that could be uncovered. For the next three hundred years scientists would attempt to reveals the rules behind the Newtonian universe.

9

EXPLORING THE
NEWTONIAN UNIVERSE

With the publication of the *Principia*, the scientific revolution that was defined by the work of Copernicus, Tycho, Kepler, Galileo and Newton had reached its apex. Yet a fundamental paradox remained between the theoretical approach to uncovering the laws of nature and the practical work of observers who spent time with their telescopes pointed at the night sky. For seventeenth century French philosopher René Descartes, who Newton had studied as a young man at Cambridge, experimental philosophy had a more ambitious agenda. Descartes was convinced that the world and universe were all made up of the same stuff, and that the laws of motion that were applicable on the surface of the Earth were also applicable at any point in the universe. These laws could thus be used to interpret the mechanistic structure of just about everything. At the heart of this philosophy was the idea that the universe was a machine that was governed by laws that could be scientifically revealed.

Newton posed a serious challenge to Descartes' mechanical universe. Where Descartes held that all motions should be explained with respect to the immediate force exerted by matter, Newton chose to describe universal motion with reference to a set of fundamental mathematical principles: his three laws of motion and the law of gravitation, which he introduced in his *Principia*. Using these principles, Newton eradicated the idea that objects followed paths determined by natural shapes (such as Kepler's idea that planets moved naturally in ellipses), and instead demonstrated that all future motions of any body could be deduced mathematically based on knowledge of their existing motion, their mass and the forces acting upon them.

Though Newton lived until 1727, his work in physics and astronomy were not as long lasting and he spent the end of his life obsessed with other non-scientific interests. Meanwhile, Newton's principles proved controversial with European philosophers, who found his omission of a metaphysical explanation for movement and gravitation philosophically unacceptable. This led to a rift between Continental and British schools of thought. The difference revolved around the question of determinism: whether mathematical laws solely governed the universe and whether everything was already determined.

Sceptics soon started to examine Newton's theories and found large anomalies in the motions of Jupiter and Saturn that were unexplained, as well as an unaccountable acceleration of the Moon's orbital speed around the Earth. The French mathematical astronomer Pierre-Simon Laplace resolved these in 1787 and in his 1805 book *Mécanique Céleste*, summarized his studies of celestial mechanics. Here he proposed that all physical phenomena in the universe could be reduced to a system of particles that exert attractive and repulsive forces on one another. He wrote:

> I have wanted to establish that the phenomena of nature reduce in the final analysis to action-at-a-distance from molecule to molecule and that the consideration of these actions ought to serve as the basis of the mathematical theory of these phenomena.

Laplace's 1796 book *Exposition du Système du Monde* summarized the general state of knowledge about astronomy and cosmology at the close of the eighteenth century. In the book, he advanced an idea that became known as the 'nebular hypothesis'. He suggested that our solar system, and indeed all stars, were created from the cooling and condensation of a massive hot rotating 'nebula' (a gassy cloud of particles). The nebular hypothesis strongly influenced scientists in the nineteenth century, and they sought to confirm or challenge it. Elements of the idea remain central to our current understanding of how the solar system was formed.

Another important philosopher in the post-Newtonian period who commented on motion in the heavens was Immanuel Kant. Born in 1724 in Königsberg, the capital of Prussia at that time, he was the fourth of eleven children of whom only four reached adulthood. Kant showed a great aptitude for study at an early age and was enrolled at the University of Königsberg at the age of sixteen (he would end up spending his entire career at this university). At university he was exposed to rationalist philosophy, and as a result became familiar with developments in British philosophy and science. It was in this context that he was introduced to the mathematical physics of Newton.

Although best known for his 1781 treatise, *Critique of Pure Reason* – now uniformly recognized as one of the greatest works in the history of philosophy – Kant made an important contribution to astronomy in 1775, when he wrote his *General Natural History and Theory of the Celestial Bodies*. In this early attempt at a theory of the heavens, he deduced that the solar system was formed from a large cloud of gas, a nebula. He further attempted to explain the order in the solar system – seen previously by Newton as being imposed by God – and correctly deduced that the Milky Way was a large disc of stars formed from a much larger spinning cloud of gas. He further suggested the possibility that other nebulae might also be similarly large and distant discs of stars. These ideas opened new horizons for astronomy, extending the field beyond the solar system to galactic and extragalactic realms, to the beginning of cosmology.

Kant pointed out that the Newtonian solar system offered a ready model for the larger stellar system where the arrangement of the stars might well be similar to that of the planets, suggesting by analogy a physical explanation for a disc structure. In other words, he suggested that the same cause that gave the planets their motion and directed their orbits into a plane could also have given the revolving stars their orbits in a plane. He called these 'island universes'.

Finally, Kant argued against the English Romantics, insisting that metaphysics could not provide an account of the foundations of physical nature, and that the issue of the existence of God was completely divorced from direct sense-experience. The Romantic Movement's distaste for science in general and Newton in particular, included some of the most revered poets and artists of the period. John Keats complained that Newton's experiments with prisms had destroyed all the poetry of the rainbow. Also, many people still today believe William Blake's famous painting of the scientist – in the form of a beautiful human body awkwardly bent over his compass – was actually celebrating Newton. In fact, the image was intended as a mockery of Newton and the idea of science.

In the late eighteenth century, after the theorizing of the great minds of Descartes, Kant and Laplace, the art of observation came to the fore of stellar astronomy in the shape of the German-English amateur astronomer, William Herschel. Born in Hanover, now Lower Saxony, in 1738, Herschel was an exact contemporary of Kant and is most well known as the man who discovered the planet Uranus. He also made many other important contributions to astronomy.

Herschel was a musician before he became an astronomer and music led him to an interest in mathematics, and thence to astronomy. At the age of seventeen years, as an oboist in the Hanoverian Guards regiment band, William was ordered to England. At the time, the crowns of England and Hanover were united under George II. This brief visit made an impression, and the following year he and his brother resigned from the Guards band and moved to London, where he quickly leaned English and settled down.

His interest in astronomy grew stronger after 1773 when he met Neville Maskelyne, who at that time was the English Astronomer Royal. Eight years later, he discovered the planet Uranus using a homemade telescope in the back garden of his house in Bath. This hobby became a kind of obsession for Herschel, during the course of his life he would construct more than four hundred telescopes.

The largest and most famous of these was a reflecting telescope with an aperture of 50 inches (1.26 metres) in diameter. On 28 August 1789, his first night of observation using this instrument, he discovered a new moon circulating the planet Saturn. The instrument was an awkward monster measuring 40 feet (12 metres) in height and at the time considered by some to be a feat of technology. This powerful telescope enabled Herschel to reach farther into space than anyone had done before, and to begin to outline the structure of our galaxy.

Herschel's sister Caroline was a great help to him during these important years and became his assistant in 1783 after he gave her one of his telescopes. Often taking notes while Herschel observed at the telescope, she soon began to make her own astronomical discoveries. She discovered eight comets, three nebulae and, at her brother's suggestion, updated and corrected the position of stars that John Flamsteed, the first Astronomer Royal during the time of Newton, had recorded. This was published as the *British Catalogue of Stars*, for which she was honoured by the Royal Astronomical Academy.

From his observations, carried out with the competent help of his sister who continued to assist him, Herschel produced the first sketch of our own galaxy. In his 1785 paper 'On the Construction of the Heavens', he wrote that the Milky Way galaxy is very extensive, a branching structure with many millions of stars. Yet Herschel's speculative cosmology failed to attract disciples. Professional astronomers could not accept his assumption that all stars are equal in brightness.

Nonetheless, the study of the nature of stars would dominate post-Newtonian astronomy and be driven by such questions as, 'is the Sun a star?' 'If so, do the other 'suns' we can see all have a moon and planets swirling around them like our Sun?' 'How far away are the stars?' And the more difficult question: 'What is the composition of the stars?'

Clearly, the first question required the application of bigger and better telescopes to look closely at the Sun. But no one had a clue where or how to start investigating the composition of the stars. Ironically the composition of the stars was eventually discovered using a surprising technique, which later became the *modus operandi* for much future work in astronomy.

Even into the eighteenth century, one hundred years after Galileo, the telescope was still unable to detect any stellar parallax caused by the Earth's motion around the Sun. Thus, estimates of the scale of the universe were pure speculation, but this did not deter astronomers from searching for results. One experiment in 1728 by James Bradley measured the aberration of light in observing the star Alpha Draconis caused by the Earth's external movement about the Sun. He set a minimum distance of 400,000 AU (Astronomical Unit – one AU is the distance between the Sun and the Earth). But astronomers wanted a more definite measurement, as some even believed that the stars could be infinitely far away!

Over one hundred years later, Friedrich Willem Bessel, a German astronomer obsessed with the accuracy of his observations, was working at his own observatory in Königsberg and managed to establish the exact positions for thousands of individual stars. He was able to observe exceedingly small motions of certain stars, relative to one another. He decided to focus on the 61 Cygni, a star barely visible to the naked eye. After correcting for its intrinsic motion, the star appeared to move in a closed elliptical path every year. This back and forth motion, called the annual parallax, could only be interpreted as being caused by the motion of Earth around the Sun. He used an unusual instrument called a *heliometer* in order to measure a nonzero angle for Cynus 61, which set the distance of 10.3 LY from the Earth to that star (LY, or light-year, is the distance a beam of light travels in one year).

This was indeed an important breakthrough, but progress using this technique was slow and frustrating. Still, parallax was the only reliable method of determining the absolute distances of nearby stars. By 1900, the parallax shift of only about 100 stars had been measured accurately and it became obvious that bigger and better telescopes were needed in order to measure the slight shift in the apparent position of stars due to the motion of the Earth. In future, more subtle methods of determining the distance to stars were developed, until by the nineteenth and twentieth centuries, sensational results were realized.

There were two important aspects to the technique of examining the faint images of far-away celestial objects. The first is the light-collecting device, the telescope; and the second is the medium of recording the image for later examination and analysis.

Invention of Photography

The search for materials to record light images began with the discovery that certain chemical compounds become insoluble when they are exposed to light. Pioneers in these methods were the Frenchman Daguerre and his contemporary, the Englishman Fox-Talbot. Daguerre discovered that iodine-treated paper was sensitive to light. By 1835, he was able to demonstrate that an invisible latent image could be produced on glass plates, developed by exposure to mercury vapour and fixed in a strong salt solution. This was the beginning of a crucial development for the future of observational astronomy.

The use of photographic equipment in conjunction with telescopes has benefited astronomers greatly, giving them two distinct advantages. First, photographic images provide a permanent record of celestial phenomena and, second, photographic plates can integrate the light from celestial sources over long periods of time. This second advantage is reliant on the rotation of the telescope being synced with the apparent rotation of the sky caused by the Earth's spin. This permits astronomers to photograph much fainter objects than they would be able to observe directly, with some exposures lasting over several days (with the shutter closed during daylight hours). John Draper of the US photographed the Moon as early as 1840 by applying the daguerreotype process and French physicists succeeded in making a photographic image of the Sun in 1845. Five years later, astronomers at Harvard Observatory took the first photographs of the stars.

Typically, the camera's photographic plate is mounted in the focal plane of the telescope. The plate consists of glass or a plastic material that is covered with a thin layer of a silver compound. The light striking the photographic medium causes the silver compound to undergo a chemical change. When processed, a negative image results. This causes the brightest spots (the Moon and the stars, for example) to appear as the darkest areas on the plate or the film.

Intuitively, the first idea for more efficient light-collecting telescopes to observe remote stars was to make the diameter of the telescope larger. The most famous example of this enhancement in the post-photography era was a 36-inch (.9-metre) diameter reflector and the historic 72-inch (183-centimetre) diameter reflector constructed by William Parsons, the third Earl of Rosse, at his home at Birr Castle in Ireland. Rosse chose the reflector design rather than the Galilean refractor because of the chromatic aberration in glass lenses but also because the construction of such a large piece of curved glass was impractical.

It was chromatic aberration that inspired Newton to design the reflecting telescope in 1668. In this configuration, a mirror reflects the rays of light on to the image point. In a Galilean-type telescope the bending, or refraction, of the light focuses the rays as it travels through the glass. However, since light of different colours is bent differently as it passes through glass, multi-images are formed, reducing sharpness. This does not occur with a mirror reflecting the incoming light onto the image plane.

Diagram of Refractor/Reflector Design

Parsons, known in history as Lord Rosse, started with basic knowledge during the 1840s, and went on to design and build the mirrors, tube and mountings for his 36-inch (.92-metre) reflector. He then built the second one to be double the size. He held the distinction of constructing the world's largest telescope in 1845. This 72-inch (183-centimetre) diameter monster held this title for three-quarters of a century. With this instrument, Rosse was able to study and record details of immensely distant stellar objects and therefore provide evidence that many otherwise mysterious nebulae were actually galaxies located far outside our own. He observed and noted various types of spirals in drawings he made of the objects that he saw, as the telescope was too unsteady for extended photographic exposures. Rosse was the first to see the spiral structure of what was later known as the Whirlpool Galaxy. He studied and named the *Crab Nebula*, based on an earlier drawing of an observation he made with his 36-inch (.9-metre) telescope that resembled a crab. A few years later, when his larger telescope was in service, he produced an improved drawing of considerably different appearance, but the original name stuck.

The Crab Nebula, a remnant of a supernova explosion, has the distinction of being the first of a list of over one hundred star clusters and nebulae compiled by the Frenchman Charles Messier in his *Catalogue of Nebulae and Star Clusters*, originally published in 1771. Consisting only of deep sky objects such as nebulae and star clusters, the purpose of the catalogue was to help fellow comet hunters to distinguish between permanent and transient objects in the sky. This catalogue certainly was useful to Rosse who listed several of the nebulae on Messier's list, including M51, also known as the Whirlpool Galaxy.

Extensive use of Rosse's telescopes was hampered, however, by the cloudy Irish weather and the fact that the metal mirror, which tarnished quickly, needed to be replaced at regular intervals. Nevertheless, the telescope and Rosse occupy an important position in the history of astronomy. In the words of Professor Sir Bernard Lovell:

> He succeeded in an almost impossible task, the measure of which can be appreciated from the fact that his telescope remained the largest in the world for three-quarters of a century. The Birr Telescope is a tribute to the third Earl's skill in engineering and optics: the results he obtained with it are a remarkable tribute to his observational skill and to his insight that such a device would record more of the depths of the universe than man had yet conceived. It is to the everlasting credit to him that he discovered the spiral structure of the nebulae and thereby opened an avenue of exploration which today has led us into the inconceivable depths of space and time.

Now it was up to the astronomers to design robust machines to allow the telescope to track the celestial images and further, to solve the problem of producing large silvered telescope mirrors. The first was a straightforward problem of mechanical design, in which the Victorians excelled. Graduated dials were mounted on the axis to permit the observer to point the telescope precisely. To track an object, the telescope's polar axis was driven smoothly by an electric motor at same speed of as the rotation of the Earth. With these innovations, one could track celestial objects with a telescope for long periods of time – and take long exposure photographs.

The next important step came from the collaboration of two specialists, namely the telescope designer and astronomer, Andrew A. Common and the mirror-maker, George Calver. They introduced several innovations which were able to produce telescopes with very smooth tracking and an unusual system that actually floated on a pool of liquid mercury.

This allowed the telescope's aspect to be continuously monitored, allowing for long, accurate exposures of the star field. Throughout his career, Common worked constantly on astrophotography and made a celebrated early photograph of the Orion Nebula in 1883, which earned him the Gold Medal of the Royal Astronomical Society.

The next advance in the construction of large-scale telescopes produced a momentous shift of international influence in observational astronomy. Englishman Edward Crossley was a businessman, Liberal Party politician and astronomer, who eventually also became a Fellow of the Royal Astronomical Society. In 1885 he commissioned Andrew Common to build him a 36-inch (.92-metre) reflector telescope, which he then housed in the astronomical observatory that he had erected. However, Crossley became so depressed by the rainy English weather and the industrial air pollution at his observatory site that he donated his telescope to Lick Observatory in California. Built to the specifications of the Common-Calver design, this instrument has since been modified but is still in use today. It is known simply as the Crossley reflector.

Crossley's bold action spearheaded a shift in leading astronomical investigation from England to the US, and the coast of California. The site that was chosen for the telescope that he donated was the peak of Mount Hamilton, situated at the southern part of the San Francisco Bay Area in northern California.

This was an excellent position where transparency and stability in the atmosphere was nearly perfect. This is important because changes in the physical characteristics of air affects the index of refraction, and will produce spurious optical effects. Furthermore, the local climate produces many clear nights for observation. When the reflector was commissioned at the turn of the century, the Lick Observatory obtained spectacular images of spiral nebulae, including the acclaimed photo of the famous Whirlpool Galaxy (classified by Messier as M51), one of the most conspicuous, and best known spiral galaxies in the sky. M51 was one of Charles Messier's original discoveries in 1773 and was sketched by Lord Rosse in 1845.

With the success of the Lick Observatory, California was now seen as the ideal location for large-aperture telescopes. The next step in the progression of improvements to observations would reinforce this for most of the twentieth century. In 1904 George Ellery Hale was recruited from the Yerkes Observatory in Chicago to act as founding director of a new observatory on the peak of Mount Wilson in the San Gabriel Mountains near Pasadena. His appointment would herald a new, epoch-making age.

As the grinding for the glass disc for the new instrument began in San Francisco, German physicist Albert Einstein published the first of a series of papers on the new theory of relativity. This theory would eventually come to describe nearly all work in astrophysics and cosmology during the twentieth century, lending this century its 'cosmic' title.

From the Laboratory to the Heavens

Pre-dating the dawn of the cosmic century was the arrival of the remarkable phenomenon of stellar spectroscopy. This technique became the perfect complement to telescopic investigations, enabling astronomers to study details of the composition and motion of the so-called fixed stars and galaxies.

A serious shift in the study of the stars took place during the eighteenth and nineteenth centuries. Astronomers were no longer content with the simple determination of the position and classification of stars, but wished to know more about their actual composition. Many were sceptical that this was an achievable goal, including the Frenchman Auguste Comte who quipped, 'Never by any means will we be able to study the chemical composition of the stars.'

In his *Cours de Philosophie Positive*, published in the 1830s, the French philosopher and sociologist Comte was expressing his pessimism about the future of astronomy, particularly with respect to the stars. As the founder of the school of positivism, which is based on natural phenomena as verified by empirical investigations, Comte thought seriously about such ideas. He used the remote position of the stars as an example of the limitation of science in explaining the universe, suggesting that the chemical composition of stars is an example of knowledge that is beyond human reach.

Few would have argued with him then. Yet, later developments proved that this prediction, like others made about science from time to time, was churlish. Before the end of the nineteenth century, more was known about the composition of the Sun and other stars than was known about our close satellite, the Moon. The make-up of the moon was only confirmed in the late twentieth century, when astronauts walked on it. Yet, we have much information about the composition of the stars. This unexpected knowledge is the result of a collaboration between physicists, chemists and astronomers from the mid-eighteenth to late-nineteenth centuries. This collaboration provided information that fuelled the great revolution in cosmology during the century that would follow.

Like much of the physics that was to come, this investigation of the stars began with the interaction of light and matter. From the time of Isaac Newton in 1664, it was known that white light contained all the colours of the rainbow. By simply passing a light beam through a glass prism, a spectrum is formed, and the white light becomes spread into different colours. In 1752 at the University of Glasgow, Thomas Melvill observed that a bright yellow light was emitted from a flame produced by burning a mixture of alcohol and sea salt. When the salt was removed, the yellow colour disappeared. The Scottish physicist then put containers of different gases over a flame and studied the glowing light that was emitted from each. He made a rather remarkable discovery: 'The spectrum of light from a hot gas when passed through a prism was completely different from the well-known rainbow light spectrum of glowing solid'. But it was not until a century and a half later that this spectrum light was analysed.

Fraunhofer Lines

It was Joseph Von Fraunhofer who eventually discovered how to apply spectra to astronomy. Fraunhofer, a German physicist who worked as an optician by trade, discovered dark lines in the spectra of stars. He noted that some of these dark lines were absent in the sun and vice versa, and took this as an indication that not all of the lines were of terrestrial origin. However, he did not describe his findings with deductive interpretations, instead confining his work to highly accurate empirical observations.

In 1814, he conducted an experiment in which he illuminated a narrow slit next to a prism and viewed the light from a glowing hot gas with a small telescope focused on the slit. Using this technique, he was able to investigate the spectrum of light coming from the gas, one narrow portion at a time. As the various colours of light have different wavelengths, he was able to calibrate his instrument to scan precisely through the entire visible spectrum between 4000 and 7000 Å. (Å is a microscopic unit of length named after the Swedish astronomer Anders Jonas Ångström. One hundred million Å = 1 centimetre.) Having created the first *spectroscope*, as it was eventually called, Fraunhofer immediately turned the instrument back on the Sun and observed a wide range of colour in its spectrum.

Fraunhofer was not surprised with the wide colour range he witnessed because he understood that the Sun is essentially a luminous white body. But he also saw in his spectroscope a countless number of sharp lines, which were darker than the rest of the background spectrum. Some of these lines appeared to be almost completely black. He reported 'an almost countless number of strong or weak vertical lines which are darker than the rest of the colour image; some appeared to be almost perfectly black'.

Though Fraunhofer was surprised by what he saw, he was convinced that these lines were due to the nature of sunlight and not an optical illusion. In 1817 he announced the following: 'I have convinced myself by numerous experiments and by various methods that these lines and bands are due to the nature of Sunlight and do not arise from diffraction or optical illusions, etc.' As Fraunhofer was the first to measure the different wavelengths of light on a numerical scale, it was natural enough that the dark lines in the solar spectrum became known as 'Fraunhofer lines'.

But why were there dark lines in the middle of the sun's spectrum? This was a puzzling phenomenon, especially since distinct bright lines were observed when pure elements like Mercury vapour (from heated liquid) and nitrogen gas were examined with the spectroscope in the laboratory. These examinations produced sharply defined and easily recognizable line patterns called spectra, which were so distinct that no two gases were seen to have the same set of bright lines. It soon became clear that line spectra, the distinguishing bright lines characteristic of a particular element, could be used to identify the presence of that element in the substance emitting the light.

Shortly after these developments, German physicist Gustav Kirchhoff took an interest in spectroscopy. In 1854, just after Melville had reported his pioneering work from Glasgow, Kirchhoff was appointed professor of physics at the University of Heidelberg, where he joined forces with Robert Bunsen (of the ubiquitous 'Bunsen burner') and founded spectrum analysis. They demonstrated that every element gives off a characteristic coloured light when heated to incandescence. This light, when separated by a prism, has a pattern of individual wavelengths that are specific for each element. Applying this new research methodology, Kirchhoff and Bunsen discovered two new elements, cesium (1860) and rubidium (1861).

Kirchhoff also applied spectrum analysis to study the composition of the Sun. He found that when light passes through a gas, the gas absorbs the same wavelengths that it would emit if it was heated. He used this principle to explain the numerous dark lines that Fraunhofer discovered in the Sun's spectrum. This would mark the beginning of a new era in astronomy.

Fraunhofer studied these dark lines using an ingenious

method. First, he made a solution containing sodium, which when coated on a hot wire, produced a pair of bright yellow lines that could easily be distinguished. Superimposing these bright yellow lines onto the Fraunhofer solar spectrum, Kirchhoff found an exact match between the bright sodium lines and some of the Sun's dark lines. This suggested to Kirchhoff that the missing lines, which now appeared dark, were due to the presence of cool sodium vapour in the outer atmosphere surrounding the Sun. Here was a radical idea that seemed to make sense.

Kirchhoff further hypothesized that when a gaseous element is extremely hot it gives off bright lines at specific wavelengths, and that these must have something to do with the internal structure of the atoms of the gas. If that same gas is cool, it absorbs light passing through it at the same distinct wavelength. In short, the cool unexcited gas absorbs light at its characteristic wavelengths, the same wavelengths at which the gas emits light when it is hot. This idea presupposes certain fundamental energy levels in the gas that are unique to that element and are reversible, that is, they can absorb or emit energy. These conclusions were a result of Kirchhoff's laboratory experiments from which he formulated three rules governing the emission and absorption of radiation by solids, liquids and gases.

- The first law is that when a solid or liquid is heated, it emits light with a continuous spectrum without any distinct bright lines.
- The second law states that if a gas or vapour composed of a single element is heated, it will emit bright lines at definite wavelengths.
- The third law states that when the same gas is cool it will absorb light at the wavelengths at which it emits light when hot, thereby reversing its characteristic spectrum from bright to dark lines.

Applying these laws to the Sun, Kirchhoff proposed that the visible surface of the Sun, or photosphere, consists of a hot liquid that emits a continuous spectrum. A cooler gaseous layer making up the atmosphere of the Sun overlays the photosphere and absorbs certain characteristic lines from the continuous spectrum. This then causes the dark lines. To support this model of the Sun's surface, Kirchhoff described an experiment that he carried out in his laboratory. He wrote:

> While engaged in research on the spectra of coloured flames, it became possible to recognize the quality and composition of complicated mixtures from the appearance of their spectra in the flame. We made some observations, which gave an unexpected explanation of the origin of the Fraunhofer lines. This allows us to draw conclusions about the composition of the Sun's atmosphere.

As he had previously done, Kirchhoff arranged his apparatus to examine the spectra of sodium vapour in his laboratory. He then cleverly superimposed the spectrum he obtained onto the spectrum of the Sun. The two bright lines from sodium, so close together that Fraunhofer labelled them with a single letter D, were easy to see. With his accurate spectroscope Kirchhoff resolved these into a pair of lines. He showed that sodium must be present in the Sun's atmosphere. He continued:

> We have noted that in the spectrum of a candle flame two bright lines occur which coincide[s] with the two dark D lines of the solar spectrum. We obtain the same bright lines in greater intensity from a flame in which common salt (sodium chloride) is introduced. We arranged a solar spectrum and allowed the Sun's rays, before they fell on the slit, to pass through a flame heavily charged with salt. When the sunlight was sufficiently weakened there appeared in place of the two dark D lines, two bright lines.

This experiment proved without a doubt that the bright lines from the sodium salt had the exact same wavelengths as the two dark D lines in the solar spectrum. He concluded that:

We may assume that the bright lines corresponding with the D lines in the spectrum of the flame always arise from the presence of sodium. The dark lines in the solar spectrum permit us to conclude that sodium is present in the Sun's atmosphere.

It seems that by studying the dark lines in the spectrum of the Sun – or even in the spectrum of a star – one could deduce what elements were present in the Sun's outer atmosphere. When, some time later, a distinct, previously unobserved pattern was discovered, a search began for the mysterious gas. The elusive element – an odourless, colourless and chemically inert gas – was finally detected and isolated in 1868. Credit for independent discovery is shared by French astronomer Pierre Janssen and English astronomer Joseph Norman Lockyer. The gas was named Helium, after the Greek word (*helios*) for the Sun.

If August Comte had been following this research, thirty-five years after his remark about 'knowledge outside the reach of man', he might feel that he had been somewhat rash. New methods of observation had quickly opened up channels to new information beyond the imagination of the French philosopher.

Spectroscopy to Study the Stars and Nebulae

This new remarkable technique, which later became known as the discipline of spectroscopy, was especially useful to astronomers in their examination of starlight. In England, even amateurs were beginning to utilize this new discipline.

William Huggins was one such English amateur astronomer. Huggins had built a private observatory at Tulse Hill in South London and in 1856, he installed a fine 8-inch (20 centimetre) refractor with optics manufactured by Alvan Clark, the world's most respected maker of refracting telescopes. Huggins went on to become a pioneer in the new technique of spectroscopy as it is applied to celestial objects. In 1864 he examined the spectrum of a planetary nebula in the constellation Draco.

He found that the spectrum had a bright line (or emission line spectrum) that took the shape of two greenish lines. These lines could not be identified as belonging to any known element at that time. Huggins assigned a name, nebulium, to the new hypothetical element.

Huggins had concluded correctly that this nebula was not composed of stars, which would have a continuous spectrum, but of glowing gas. He was therefore the first to observe the spectrum of a planetary nebula. As he wrote:

> On the evening of the 29th of August, 1864, I directed the telescope for the first time to a planetary nebula in Draco. The reader may now be able to picture to himself to some extent the feeling of excited suspense, mingled with a degree of awe, with which, after a few moments of hesitation, I put my eye to the spectroscope. Was I not about to look into a secret place of creation? I looked into the spectroscope. No spectrum such as I expected! A single bright line only!

Subsequent observations, with better resolution, showed that this bright line was actually a double line. Huggins showed that nebulae had bright emission lines, unlike the broad spectrum expected of unresolved stars of which nebulae where widely believed to be composed. He concluded that nebulae are enormous masses of hot luminous gas or vapour, which would never resolve into stars.

On 18 May 1866, Huggins made the first spectroscopic observation of a nova and found emission lines of hydrogen. In 1868 he took the spectrum of a comet and identified the spectral lines of ethylene. In later life he was also helped by his wife Margaret, whom he married in 1875. Margaret was extremely competent and helped Huggins carried out extensive observations of the spectral emission lines and absorption lines of various celestial objects. They were the first to distinguish between nebulas and galaxies by showing that some like the Orion Nebula had emission spectra characteristic of a gas,

while others like the Andromeda Galaxy had spectra characteristic of stars.

This was the beginning of a new age in astronomy, an age in
which spectroscopy would lead the way to a more complete
examination of the universe: stars, nebulae and galaxies.

This new age would adopt the name astrophysics, and
become a field that classified stars in the same way that plants
and animals were grouped to enhance understanding of what
makes them different. One of the starting points for this new
field grew out of the interesting career of Henry Draper, an
American and one of the pioneers of the use of astrophotography. In 1872 Draper photographed the first stellar spectrum
that showed absorption lines. On 30 September 1880, he also
became the first to photograph the Orion Nebula.

Draper had ideal parentage for an aspiring astro-photogra-
pher. His father, John William Draper, was an accomplished
doctor, chemist, botanist and professor at New York University.
In the winter of 1839–40 he also became the first to photograph
the moon through a telescope. Draper's mother was Antonia
Coetana de Paiva Pereira Gardner, daughter of the personal
physician to the Emperor of Brazil. After graduating from the
New York University Medical School at the age of twenty,
Draper visited the colourful Lord Rosse in Ireland. There,
under the spell of the eccentric Irishman, he was inspired to
consider the application of photography to astronomical
observation. He produced many innovations in his new chosen
field, especially improvements to the photographic process.

In 1867 he married Anna Mary Palmer, a wealthy socialite
who, upon his untimely death at age forty-five, established
the Henry Draper Fund at Harvard Observatory. These
funds were used to prepare the Henry Draper Catalogue of
stellar spectra under the supervision of Edward Pickering, a
dynamic and committed physicist. Pickering's plan was to
mount a large systematic study of the stars, incorporating the
continued development of the telescope as well as the improvement of astrophotography. But this required a substantial

concentration of resources. Fortunately, the Harvard College Observatory was generously endowed and about to become the world's foremost centre for stellar spectroscopy.

The Draper Memorial Catalogue describes the spectra and properties of over 10,000 stars. Published in 1890, it provides the largest and most systematic classification completed during the nineteenth century. The establishment of the Harvard College Observatory at the close of the nineteenth century is noteworthy on two counts: First, Pickering proved to be an inspired, almost obsessive leader in the pursuit of accurate stellar data and classification. He often carried out observations himself and contributed thousands of dollars of his own money to run the observatory.

Second, Pickering was certain that women had a part to play in the story of astronomy, especially in the difficult task of recording and classifying stars. For this task he hired Williamina Fleming, Antonia Maury and Annie Cannon, all of whom joined the staff of HCO in 1896. A large team of women 'computers' supported these four. Annie Cannon was reportedly responsible for single-handedly classifying almost 400,000 spectra in her nearly fifty years at HCO. During one period of only four years, Cannon classified some 225,000 stars. For this she received many international honours, including an honorary degree from Oxford University.

After Pickering died in 1919, he was succeeded by Harlow Shapley and a new era at the Harvard College Observatory.

PART III

EINSTEIN'S UNIVERSE

10

RELATIVITY: SPECIAL AND GENERAL

At the beginning of the twentieth century, two major developments were beginning to dramatically expand the field of cosmology. The first of these was the improvement in design and construction of large telescopes; the second was the development of a theory by a German scientist, Albert Einstein, which astonishingly could be applied to the universe as a whole.

Technical improvements in the construction and use of telescopes was mainly an American industry led by entrepreneurs such as George Ellery Hale (Chicago), who built the first two large observatories in the US, Harlow Shapley (Harvard) and observers Vesto Slipher (Lowell) and Edwin Hubble (Mount Wilson). On the other hand, Europeans such as Einstein (Germany), Arthur Eddington (Cambridge), Willem de Sitter (Holland) and Georges Lemaître (Belgium) led the theoretical advances.

Today, research labs try to maintain a mix of experimentalists and theorists in order to enhance the interplay between technical and theoretical advances. This was not possible in the 1920s when state-of-the-art theoretical work was being carried out in Europe and the giant telescopes were being built in America. Ironically, these years were perhaps the most important in modern history, for it was during this time that astronomy – the scientific study of celestial objects such as stars, planets, comets, galaxies and phenomena that originate outside the Earth's atmosphere – developed into cosmology, the study of the universe in its totality, and humanity's place in it.

It is perhaps best to separate these two developments, the theoretical and the experimental, in order to trace an uninterrupted understanding of their respective chronologies. Before describing the groundbreaking advances that occurred with respect to equipment, techniques and new methods for obtaining astronomical observations, the revolutionary theoretical work of Einstein and his followers will be reviewed.

1900 and the New Age of Physics

At the turn of the century, classical physicists believed that most of the important theoretical work in their subject had been successfully completed. Applying the Newtonian mechanical model, nineteenth century scientists like Michael Faraday and James Clerk Maxwell applied careful experimentation and elegant equations to explain nearly all electricity and magnetism phenomena. Maxwell's equations showed that light itself was a propagation of electromagnetic waves travelling at a speed, which could be computed independently from measurable physical constants. Furthermore, the statistical methods of Ludwig Boltzmann described the behaviour of gases. This led to the viable prospect of explaining all of physics using classical concepts.

This period could be summed up by one of its most accomplished experimentalists, Albert Michelson. After using the wave properties of light to quite accurately measure its speed, Michelson believed that only trivial problems remained to be solved:

The most important fundamental laws and facts of physical science have all been discovered, and these are now so firmly established that the possibility of their ever being supplemented in consequence of new discoveries is exceedingly remote.

This remark has become famous for its naiveté, though it did represent the general feeling of most physicists at the time. However, this was not the case for a young man daydreaming through his work as a clerk in the Swiss patent office in Berne. Albert Einstein had some quite fundamental conceptual concerns with the core of Newton's classical structure – the structure that was otherwise accepted by most other scientists.

To appreciate the depth of Einstein's unease, imagine Newton's theoretical structure as a house of cards where each layer is built on the one below. Einstein only removed two of these cards – Newton's concept of universal time and his concept of absolute space – but they just happened to be at the base of the structure.

In order to debunk these two concepts, it was necessary to postulate that nothing could travel faster than the speed of light, which Einstein said was seen to have the same value to all observers, moving or stationary. This work he called the 'special theory of relativity', which he described in one of the three papers he published in 1905.

Einstein's first papers of that year were about electrodynamics and took as their concern the relationship between light signals and moving clocks. This led him to engage with gravitation and he became troubled by its bewildering property of action at a distance. According to Newton, if the Sun disappeared in an instant, so would its gravitational field on the surface of the Earth, millions of miles away. Newton's formulation didn't bear any consideration of the speed of transmission of the gravitational field, which in this example would be infinite. We know, however, that if the Sun disappeared, its light, with its finite speed, would continue to travel towards the earth for another eight minutes. This discrepancy troubled Einstein.

So did the concept of mass. There appeared to be no reason why inertial mass, which is a measure of a body's resistance to motion, and gravitational mass, which is a measure of a body's pull on other bodies, should be exactly equal. Einstein pondered these ideas for years, and eventually began to consider if there might be another way to explain gravity. Maybe gravity was not a force at all. Since the motion of a free-falling object does not depend on the object's mass (as Galileo discovered in the sixteenth century), gravitation might be due to certain properties of the medium it's falling through, that is, space itself.

Through a series of remarkably creative and idiosyncratic steps, Einstein then decided that space was not flat, but curved. The local curvature, he theorized, was produced by the presence of mass in the universe. Consequently, bodies near other massive bodies do not travel in straight lines. Rather, these follow the path of least resistance along contours of curved space (called geodesics) that are themselves caused by the nearby massive bodies. He knew that if this were true, there would be no need for a mysterious 'force of gravity', which is transmitted instantaneously. Nor would it be necessary to explain the odd coincidence that inertial and gravitational mass are exactly equal. Einstein set out to rescue classical physics from these inconsistencies and thus finish the task started by Galileo, Newton, Faraday and Maxwell; in the process, he discovered a whole new universe.

Most great works in physics have come from those who combine miraculous physical intuition with sound mathematical skills. The former is far more important than the latter, and great physicists like Einstein don't worry too much if they don't have the necessary mathematical formalism at hand when they discover new physical principles. In 1912, when Einstein realized his idea of curved space had some plausibility but required a non-Euclidian geometry, he was not clear how to quantify his new approach. He therefore contacted his school friend Marcel Grossmann, who suggested he learn the techniques of Riemann geometry, the most general transformation formulae known for frames of reference of the type Einstein seemed to require. The two friends struggled for a few years, and finally achieved the formalism that was necessary to support general relativity in 1915. Einstein published his famous paper on the theory in the following year.

During this period Einstein fashioned the form of the equations by dreaming up more thought experiments (*gedanken*), much as he had done when he composed his theory of special relativity ten years earlier. He wrote of this process, which he called 'the happiest thought of my life':

Sitting in a chair in the patent office in Berne (in 1907), a sudden thought occurred to me. If a person falls freely he will not feel his own weight. I was startled and the simple thought made a deep impression on me. It impelled me towards a theory of gravitation and was the happiest thought of my life. I realized that for [an] observer falling freely from say, the roof of a house, there exists – at least in his immediate surroundings – no gravitational field. If the one who is falling drops other bodies [like Galileo's cannonballs], than these remain relative to him in his state of rest or of uniform motion independent of their particular chemical or physical nature. The observer has the right to interpret his state as at rest or in uniform motion . . .

He continued:

Because of this idea, the uncommonly peculiar experimental law that in the gravitational field all bodies fall with the same acceleration attained at once a physical meaning. [This is the same as saying that gravitational mass is the same as inertial mass.] If there were to exist just one single object that falls in a different way than all others, then with its help the observer could realize that he is in a gravitational field and is falling in it. However if such an object does NOT exist – as experience has shown with great accuracy starting with Galileo in 1590 – then the observer lacks any objective means of receiving himself as falling in a gravitational field. He has a right to consider the state as one of rest and his environment as free of gravity. Therefore, the fact that the acceleration of free fall is independent of the nature of the material involved is a powerful argument that the relativity postulates can be extended to coordinate systems which are in non-uniform motion.

Einstein's thought that 'a person falling freely does not feel his own weight' seems rather simple. Yet from this starting point, he squeezed every possible drop of insight, removing all the inconsistencies of Newton's theory. He transformed a simple picture of someone falling through space into a small laboratory in which he proved that gravity did not exist. He could then analyse the effect of gravity on such phenomena as the bending of light beam or the slowing of a clock by simply replacing the gravitational field with simulated accelerated motion.

By simply thinking about a man jumping off a roof in Berlin (or so the story goes), Einstein was able to replace gravity by acceleration and discover his principle of equivalence. He had noticed that it was not possible in a closed system, i.e. one in which observations are only possible within the system, to distinguish between the effects of gravitation on a body and the acceleration of that body. He deduced that they are equivalent and can be used interchangeably. Later, his sketchy qualitative ideas about curved space and his application of the equivalence principle matured into a set of equations which gave the precise amount of curvature for a given quantity of mass. This development is considered as one of the most creative examples of the power of pure abstract thought.

Einstein could now use the powerful principle of relativity – that the laws of physics should not depend on any particular reference frame – to test his new laws of space curvature. He used the principle of equivalence (gravity equals acceleration) as his starting point. He had one more useful bit of information, a single observation: the anomalous advance of the perihelion of the planet Mercury. Astronomers had known for years that the farthest point of the orbit of Mercury, its perihelion, moved slowly around the ecliptic over time and did not return to the same starting point at the end of each cycle. No one could explain this effect using classical theory.

Seventeenth century scientists were not worried about this small discrepancy in Mercury's elliptical orbit. However, by Einstein's time, astronomers were concerned, and needed an explanation. The discrepancy had later been carefully measured to be 43 seconds of arc per century and the problem could not be ignored. Einstein saw that he could now use the perihelion results to test his curvature law.

He first produced a set of equations consistent with his new postulates and then proceeded to test them. He found that his equations gave the correct prediction for the anomalous shift of Mercury's perihelion. Calculations using the Newton theory predicted an advance of the perihelion point of about 531 arc-seconds per century. However, careful measurements using the world's best telescopes indicated that the actual advance was 574 arc-seconds per century, a difference of 43 arc-seconds. It was no doubt satisfying for Einstein to discover that a calculation using general relativity accounted for the correct value of the remaining discrepancy, 43 arc-seconds per century.

Importantly, the equations also incorporated the equivalence principle and obeyed the principle of relativity. That is, the equations have the same form when expressed in each and every reference frame that Einstein could apply. The final form of his equations also predicted the deflection of 1.7 seconds of arc for starlight passing near the edge of the Sun and incorporated his earlier prediction of gravitational time dilation. This is today referred to as gravitational red shift and refers to the phenomenon in which clocks placed in a strong gravitational field would slow down. Confident with his findings, Einstein presented the final form of his general relativity law of curved space and warped time to the Prussian Academy on 25 November 1915.

Soon after, the elated scientist sat down and wrote a letter to a close friend, the Dutch physicist Paul Ehrenhest:

Dear Ehrenhest,
I have been beside myself with ecstasy for days. Imagine my joy that the new law of curvature obeys the principle of relativity

and predicts the correct perihelion motion of Mercury. I can tell you that the days of searching in the dark for a truth that one feels he cannot express – the intense desire of the alterations of confidence and misgivings until one breaks through to clarity and understanding – are known only to him who has experienced them himself.

The thirty-six-year-old physicist had produced a set of mathematical equations that gave details of the relationship between the curvature of space and the distribution of mass in the universe. Simply put, Einstein found *that matter tells space how to curve and then space tells matter how to move*. This was a new way to describe gravitation – without forces. A complete change of perspective was necessary to jump between the two pictures of gravitation. Contained within these miraculous equations was the correct prediction of perihelion shift of Mercury; the degree of bending of starlight; the existence of gravitational waves; information about the singularities of space time; the description of the formation of neutron stars and black holes; and even the prediction of the expansion of the universe, which was at that time yet to be discovered.

The bad news was that the mathematics was extremely difficult as there were some twenty simultaneous equations with ten unknown quantities. The equations are almost impossible to solve except in situations where symmetry or energy considerations reduce them to simpler forms. One solution proposed in 1917 by the Dutch cosmologist Willem de Sitter became particularly important despite it being predicated on the absurd assumption that there is no matter in the universe. De Sitter, appointed to the chair of astronomy at Leiden University in 1908, was director of the Observatory from 1919 until his death in 1934. He was quite important in the early application of Einstein's new theory, especially in Europe. He had studied and worked at the University of Groningen and the Cape Observatory in South Africa before taking his post at Leiden.

The de Sitter solution bothered Einstein since he believed the presence of matter throughout the universe was necessary to support the proper inertial structure of general relativity's space-time. To ensure this, Einstein had arbitrarily incorporated a 'cosmological constant', indicated by the Greek letter lambda, into his equations. This guaranteed the static distribution of stars required by the reports of astronomers. Many accepted de Sitter's empty universe solution, arguing that the density of matter is so small that a 'zero-density' solution was perhaps a good first approximation.

Einstein's set of equations is written below with the various terms identified. Note that if the cosmological constant and the mass density are taken as zero (empty universe) the vacuum solution is obtained.

Einstein's cosmological constant (lambda)

$$R_{ik} - \Lambda g_{ik} = 8\pi T_{ik}$$

Metric tensor

Mass density (source of curvature)

Einstein explains his field equations.

In 1916 de Sitter and Ehrenhest suggested to Einstein that a spherical form of his new four-dimensional space-time would eliminate the problems of the boundary conditions at infinity, which had posed impossible difficulties for Newtonian cosmological models. As a result of this suggestion, Einstein realized that in his theory of general relativity he had for the first time a theory that could be used to construct models for the universe as a whole. This is often considered to mark the beginning of theoretical cosmology.

Applying relativistic field equations in this way constituted a revolution in traditional conceptualizations of the universe. Einstein had essentially redefined the idea of cosmology with his new theory. The universe had previously been thought of as those parts that existed within the limits of observation. This was dependent on the range of telescopes that existed at the time, near the end of the nineteenth century, and these telescopes were not entirely precise.

Einstein ultimately conceptualized the universe as the totality of events in space and time – and put forth one set of equations to describe this totality. Though the change was accepted by only a small number of physicists, mathematicians and astronomers at the time, the great significance of this new relativistic cosmology became evident in retrospect. But Einstein and his followers decided that if cosmology was to progress and become a science, certain objections to general relativity had to be ignored for the time being. And so the development of the new theory continued.

In order to understand and elaborate on the two relativistic world models of Einstein and de Sitter, some scientists proposed solutions that combined features of both models. With two notable exceptions in the 1920s, the framework was essentially confined to these static world models, even though the tendency was to believe that neither of these classical solutions represented the actual universe.

De Sitter's connection with Leiden became important immediately after the publication of Einstein's theory of general relativity because Holland had remained neutral during the First World War (1914–18) and it was therefore still possible for a German national to travel there. Ehrenhest was a close friend of Einstein's, and would often invite him to visit the university to discuss important matters in physics. However, Einstein's relationship with de Sitter, who was a cosmologist and head of the astronomy department, was more critical.

De Sitter's Dutch nationality is therefore an important detail. Neutral Holland was the only immediate conduit for the transmission of Einstein's new theory to physicists outside Germany. The Dutchman was also a foreign member of the Royal Astronomical Society (RAS), where he frequently contributed papers and discussion. During the First World War Arthur Eddington, Britain's most important astronomer, was Secretary of the RAS. Eddington was therefore the first person to see de Sitter's series of letters and papers regarding Einstein's theory. The Englishman was fortunate in this regard for two reasons: firstly, he was one of the few astronomers with the mathematical skills to understand general relativity; and secondly, owing to his international and pacifist views, one of the few Englishmen who was willing to pursue a theory developed by a German physicist. Eddington quickly became the chief supporter and expositor of relativity in Britain.

Arthur Eddington

Arthur Eddington was born in Kendal, England, near the Lake District. The son of devout Quaker parents, his father taught at a Quaker training college in Lancashire before moving to Kendal to become headmaster of Stramongate School. When his father died in the typhoid epidemic that swept England in 1884, the family moved to Weston-super-Mare where Arthur was first educated at home before spending three years at a preparatory school.

From an early age, Eddington was recognized as having a brilliant mind, a strong work ethic and a steely determination. In school, he proved to be a very capable scholar particularly in mathematics and English literature, and in 1898, at the age of sixteen, earned a scholarship to Owens College, Manchester. Here he spent the first year in a general course, but turned to physics for the next three years. Eddington was greatly influenced by his physics and mathematics teachers and came under the lasting influence of the Quaker mathematician J.W. Graham. His progress was rapid, winning him several scholarships, and he graduated with a BSc in physics with First Class Honours in 1902.

He was awarded a scholarship to the University of Cambridge by Trinity College in 1902, where his tutor, R.A. Herman, was another distinguished mathematician. Eddington came of age at just the right time for the new revolution in mathematical physics, receiving his BA in 1905, the year of Einstein's first paper on relativity. His rise in the ranks of English astronomers would be meteoric.

In January 1906, at the age of twenty-four years, Eddington was nominated to the post of chief assistant to the Astronomer Royal at the Royal Greenwich Observatory. He left Cambridge for Greenwich the following month, was put to work on several challenging projects – which he dispatched with distinction – and in no time had won a Fellowship at Trinity.

In December 1912 George Darwin, son of Charles Darwin, died suddenly and Eddington was summarily promoted to his chair as the Plumian Professor of Astronomy and Experimental Philosophy at Cambridge. Later in 1913, Robert Ball, holder of the theoretical Lowndean chair also died, and Eddington was named the director of the entire Cambridge Observatory at only thirty-one years of age. He was elected a Fellow of the Royal Society shortly after.

During the First World War, Eddington became embroiled in controversy within the British astronomical and scientific communities. As a Quaker pacifist, he struggled to keep his wartime politics separate from astronomy and repeatedly called for British scientists to preserve their pre-war friendships and collegiality with German scientists. When called up for conscription in 1918, he claimed conscientious objector status, a position recognized by the law if somewhat despised by the public. The government sought to revoke this deferment and only the timely intervention of the Astronomer Royal and other high profile figures kept Eddington out of prison. This was very fortunate for the world of science as he was just about to perform an experiment that would confirm the validity of Einstein's new, bizarre theory of the universe.

A Test for the New Theory of Gravitation

Einstein had spent ten years of strenuous work on his new theory of gravitation. He hoped it would replace that of Isaac Newton's, but knew that under most circumstances, for example when masses are not large, the old theory provided the right answers. Though the new theory importantly removed some of the problems of Newton's formulation, like the instantaneous transmission of the gravitational field and the odd behaviour of the planet Mercury's motion, it also postulated a whole new mechanism.

Gravity was not caused by the attraction of one body for another, said Einstein, but fundamentally by the distortion of space by mass. Since the distortion of space predicted by the theory was such a small effect, Einstein knew it would be almost impossible to test on earth. Only large masses, like the planet Jupiter or the Sun, would produce effects that could be detected. One of the predictions of the theory was that a spaceship in inter-galactic space could, for example, fire a laser beam just past a massive planet while another spaceship cruised by to see if the light was bent by the planet. It has always been supposed that light travels in a straight line through space. But if space was curved as Einstein predicted, the light would follow the curvature. This could give a detectable deviation of the path of the light beam and thus a way to test the new theory. It was true that the initial successes of the theory had been impressive, but a more direct test would be necessary to satisfy critics before they would accept the overthrow of arguably the most important discovery in scientific history, Newton's universal gravitation. But how to devise such a test in the real world?

Einstein, who is usually thought of as otherworldly and completely theoretical in the fields of mathematics and geometry, knew the answer. He suggested a remarkable astronomical experiment, which could only take place during a total eclipse of the sun. He described such a test in the last paragraph of his first paper on general relativity: 'As the fixed stars in the parts of the sky near the sun are visible during a total eclipse of the Sun, the consequence of the theory may be compared with experiment.'

As soon as the theory was announced, a colleague of Einstein's, Erwin Freundlich of the Potsdam Observatory near Berlin, examined photographs of past eclipses but found them to be inconclusive. The Lick Observatory in the US then tried to photograph the solar eclipse of 8 June 1918, but was defeated by cloudy skies. So when the war ended with the armistice of November 1918, the British became the next to take up the challenge. And who better to try the experiment than the man who had promoted Einstein's theory in the English-speaking world, Arthur Stanley Eddington?

The date of the first total solar eclipse to occur after the war was already known. However, a large telescope and high-resolution camera then had to be transported into the path of totality and set up to photograph the eclipsed sun. When second contact occurred – when the moon completely covered the sun – the stars and planets beyond would appear in the darkened sky. The light travelling from certain bright stars in the background field of the Sun, travelling from unimaginable distances for millions of years and passing near the outer rim of the eclipsed Sun, would enter the camera and expose the photographic plate.

If Einstein was right, the beams passing close to the Sun would be bent by the curvature of space caused by the large Sun's mass and would be deviated to a new path before reaching the camera on earth eight minutes later. To detect the shift

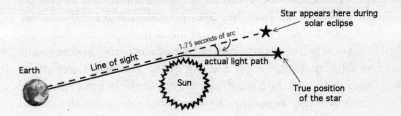

Bending of starlight by the Sun during a solar eclipse.

in the apparent position of the stars, it would be necessary to compare the photos of the displaced positions with photos taken of the same field of stars months earlier, when the Sun was in a different position along the ecliptic and the starlight was undeviated.

For this experiment to succeed the sky would have to be clear; the resolution of the camera needed to be high enough to produce sharply-focused star images on the film plates and the stars had to be bright enough to be visible in the vicinity of the eclipsed Sun. There is not much that experimenters could do about the vagaries of the weather. But they could make certain the camera was capable of taking high resolution images and select an eclipse with a bright enough star field background to photograph. This last requirement may have been a problem because there are only about two eclipses each year.

Given the rarity of a total solar eclipse and the scarcity of bright stars along the Sun's annual path, astronomers might have waited years for such favourable circumstances. But in this case, they didn't have long to wait. If the astronomers planning such a project had had their choice in selecting the optimum star field for such an eclipse photo experiment, they would probably have chosen the day of the year when the Sun, in its journey through the constellations of the Zodiac, was positioned in the middle of the bright star cluster called the Hyades. In the present epoch, the Sun is in Hyades on the 29 May each year. Imagine the enthusiasm for the eclipse test coordinators when it was realized that the next total solar eclipse – the first since the ending of the war – was to occur on 29 May 1919, the precise day of the year when conditions were optimum. Furthermore, this particular eclipse was to be 6 minutes and 50 seconds in duration, close to the theoretical maximum.

So it was not surprising that even while the war was still in progress, Sir Frank Dyson, the Astronomer Royal, had seized the opportunity afforded by this particular eclipse to test Einstein's theory. He obtained a Government grant of one thousand pounds and for help turned to Eddington, a long-standing member of the Royal Observatory staff as well as translator and expositor of the difficult relativity theory.

It has generally been understood that as the leading proponent of Einstein's new theory and one of the world's top astronomers, Eddington was the driving force behind the eclipse expedition, but this is not true. For one thing, he had no doubt that Einstein's theory was correct and did not feel that it was necessary to confirm the predictions with an experiment as difficult as photographing a solar eclipse. Had not the theory already explained the motion of the perihelion of the planet Mercury that Newton's theory failed to do? Eddington thus became involved in the eclipse project as a kind of 'punishment' for his wartime pacifist beliefs.

In order to properly execute the experiment, preparations would have to begin about two years in advance. Although Dyson did have a small grant, it was impossible to ask any instrument maker for help when all technical expertise was required in the war effort. And there were other complications as well. As an able-bodied thirty-four-year-old male, Eddington was subject to the draft. But as a devout Quaker, it was well-known that he would claim deferment as a conscientious objector to war and could end up peeling potatoes in a camp in northern England with other like-minded pacifists.

The Cambridge big shots argued effectively with the authorities that it was not in the country's best interest to have as distinguished a scientist as Eddington serve in the army. They no doubt used the sad case of the brilliant young crystallographer William Mosely, killed at Gallipoli in the First World War, to convince them. Finally, Eddington received a letter of deferment from the Home Office; all he had to do was sign and return it. But Eddington stubbornly added a postscript stating that if he had not been deferred, he would have claimed deferment as a conscientious objector. Much to the frustration of the Cambridge crowd, the Home Office was furious and ready to send Eddington away with his Quaker friends; but the Astronomer Royal Dyson intervened directly. In the end, Eddington was deferred with a stipulation that if the war ended before May 1919, he would undertake to lead an eclipse expedition to verify Einstein's theory.

A joint committee of the Royal Society and the Royal Astronomical Society led by the Astronomer Royal was set up to organize the expedition. They planned two separate experiments, one in Sobral, Brazil and the other on a small island Príncipe off the west coast of Africa. As the war ended, Eddington confided to Dyson that he had an intense spiritual feeling about the project; and that on these grounds he decided to travel to Príncipe. He set sail for Lisbon in March 1919.

When Einstein calculated the angle through which a light beam would bend if it grazed the edge of the Sun, he predicted 1.75 seconds of arc, exactly twice the amount calculated previously using Newton's theory. This led him to suggest that the results of a precise eclipse photographic study would support either his own theory of gravitation or that of Isaac Newton. The eclipse measurement would have to be carried out carefully to distinguish between Einstein's prediction and the Newton result of half that value. The latter had been calculated as early as 1804 by assuming that the Sun's gravitational field would deflect light corpuscles passing the Sun's edge. Though the two men's values differed by only a factor of 2, their calculations are based on completely different physical principles. The measurement would constitute a critical test.

For weeks on the island Eddington fussed with the telescopic mount, which apparently was acting erratically. He built special observing huts and finally waited for the full Moon's path to cross the ecliptic. Rain fell from 10 May until the morning of the eclipse, and as the moment for the eclipse approached on 29 May, the Sun was completely hidden behind the clouds. Finally, praying for a miracle, Eddington despairingly pointed his telescope to the Sun and removed the shield covering the photographic plate. As the eclipse occurred, he removed the lens cap from the telescope and took several photographs of the dark cloudy sky. The 400 seconds of totality ticked by and Eddington could see no stars in the view. Miraculously, the clouds suddenly began to evaporate and a few stars were visible for just a moment. He quickly exposed a fresh plate before the Sun appeared again from behind the Moon.

He developed the plates and found star images on one single plate. Eagerly he made tentative micrometer measurements and to his delight found a displacement of 1.61 plus or minus 0.3 seconds of arc – results which favoured Einstein's new theory over Newton's. Eddington's notebooks reveal his personal account of the expedition:

> We got our first sight of Príncipe in the morning of April 23 and soon found we were in clover, everyone anxious to give every help we needed. About May 16 we had no difficulty in getting the check photographs on three different nights. I had a good deal of work measuring these. On May 29 a tremendous rainstorm came on. The rain stopped about noon and about 1:30 p.m., when the partial phase was well advanced, we began to get a glimpse of the sun. We had to carry out our programme of photographs in faith. I did not see the eclipse, being too busy changing plates, except for one glance to make sure it had begun and another half-way through to see how much cloud there was. We took 16 photographs. They are all good of the Sun, showing a very remarkable prominence; but the cloud has interfered with

the star images. The last six photographs show a few images, which I hope, will give us what we need . . .

A few days later there was another entry:

> June 3. We developed the photographs, 2 each night for 6 nights after the eclipse, and I spent the whole day measuring. The cloudy weather upset my plans and I had to treat the measures in a different way from what I intended, consequently I have not been able to make any preliminary announcement of the result. But the one plate that I measured gave a result agreeing with Einstein.

This was an instant that Eddington never forgot, later referring to it as the 'greatest moment of his life'.

Upon returning to England, the eclipse photographs were analysed very carefully before the results were finally communicated. By September rumours had reached Einstein that the eclipse results were favourable. Even though the war was now over, Eddington was still communicating through his colleagues in Holland. On 22 September, Einstein received a telegram from his friend, the respected Dutch physicist H.A. Lorentz, that Eddington had found the full deflection of the starlight in agreement with general relativity. Einstein immediately sent a postcard to his ailing mother in Switzerland: 'Dear Mother, Good news today. H.A. Lorentz has wired me that the British expedition has actually proved the light deflection near the Sun.'

It would appear from this postcard that Einstein had been slightly uncertain how the eclipse result would turn out. That being said, he was never uncertain about his theory, as is suggested by another story from this period which has been since repeated in the common rooms of physics and astronomy departments the world over.

In 1919 a student of Einstein's at the University of Berlin, Ilse Rosenthal-Schneider, reported that one day when she was studying with the professor, he suddenly interrupted the discussion and reached for a telegram that was on the window-sill. He handed it to her saying, 'here, this will perhaps interest you.' It was a cable from Eddington with the results of the eclipse measurements. After she expressed her joy that the results coincided with his calculations, he said, quite unmoved, 'But I knew that the theory is correct'. This prompted her to then enquire about how Einstein would have responded if there had been no confirmation of his prediction. 'Then I would have been sorry for the dear Lord', he responded, 'because the theory is correct'.

Einstein replied shortly after to his English colleague, addressing him excitedly with 'Lieber Herr Eddington!':

> Above all I should like to congratulate you on the success of your difficult expedition. Considering the great interest you have taken in the theory of relativity even in earlier days I think I can assume that we are indebted primarily to your initiative for the fact that these expeditions could take place. I am amazed at the interest, which my English colleagues have taken in the theory in spite of its difficulty.

But the news was still not official. On 6 November 1919, an historic joint meeting of the Royal Society and the Royal Astronomical Society was held in London. More than two centuries earlier, Newton had been elected President of the Royal Society, and annually thereafter he was re-elected until his death. In 1919, he was vividly present in the minds of the assembled scientists; his portrait held pride of place on the wall, dominating the scene. Yet though he faced the audience, his eyes seemed to be turned sharply to the right as the Astronomer Royal officially announced:

> Thus, the results of the expedition can leave little doubt that a deflection of light takes place in the neighbourhood of the Sun and that it is of the amount demanded by Einstein's general theory of relativity.

As the meeting came to an end, Joseph John Thomson, who discovered the electron, but who was also a Nobel Laureate and President of the Royal Society, publicly hailed Einstein's work as 'one of the greatest – perhaps the greatest – achievements in the history of human thought.' The drama of the occasion was undoubtedly heightened by the war that had just ended. On a small African island, an English Quaker astronomer had confirmed a new theory of the universe, the brainchild of a German Jew working in Berlin.

The predicted deflection had been confirmed under circumstances of high drama, at a time when nations were war-weary and heartsick. The bent rays of starlight had brightened a world in shadow, revealing a unity of man that transcended war. But fate had played an unexpected trick. The deflected starlight had dazzled the public and suddenly Einstein was world famous. This essentially simple man, a cloistered seeker of cosmic beauty, was now a world symbol and the focus of widespread public adoration.

The attention that his discovery garnered started with the headlines in the London *Times* in an article headed 'The Fabric of the Universe', which explained the Eddington expeditions and their purpose. The article concluded that:

> It is confidently believed by the greatest experts that enough has been done to overthrow the certainty of ages, and to require a new philosophy, a philosophy that will sweep away nearly all that has been hitherto accepted as the axiomatic basis of physical thought.

The Times was not the only major newspaper to make so much of relativity theory and the Eddington expedition. The *New York Times* of 9 November 1919, carried a three-column article headed, 'Eclipse Showed Gravity Variation: Hailed as Epoch Making.' After describing the expeditions, the paper stated that 'The evidence in favour of the gravitational bending of light was overwhelming, and there was a decidedly stronger case for the Einstein shift than for the Newtonian one.' The article then quoted J.J. Thomson's assessment of the experimental verification of relativity. 'It is not a discovery of an outlying island, but of a whole continent of new scientific ideas of the greatest importance to some of the most fundamental questions connected with physics.'

Perhaps we should let Eddington have the last word. In a Memorial lecture on Eddington several years ago, Subrahmanyan Chandrasekhar confirmed an anecdote, which for many years was thought to be apocryphal. As the 1919 joint meeting of the Royal Society and the Royal Astronomical Society was breaking up, a reporter approached Eddington and commented, 'Professor Eddington, I hear that you are one of only three persons in the world who understand general relativity'. Eddington, somewhat bemused, scratched his head. 'Don't be modest, Eddington', the reporter said. To which the unassuming professor replied, 'Oh no, on the contrary, I am trying to think who the third person is'.

At a later banquet held in London in 1930 to honour Einstein, the Master of Ceremonies George Bernard Shaw, made the following statement, much to the delight of the guest of honour:

> Ptolemy made a universe, which lasted fourteen hundred years. Newton also made a universe, which has lasted three hundred years. Einstein has made a universe, and I can't tell you how long that will last.

Alexander Friedman

Whether the eclipse results and the astounding worldwide publicity that followed had any effect on the subsequent development of the new theory is hard to say. What can be confirmed is that the eclipse results provoked a renewed effort, mostly in Europe, to find solutions to Einstein's miraculous equations. The first general relativistic models, including that of Einstein himself, predicted a universe that was dynamic and contained ordinary gravitational matter, a universe that would contract or expand.

Einstein's first proposal for a solution to this problem involved adding the artificial term 'lambda', the cosmological constant, into his field equations. This functioned to essentially stop the expansion. During this time astronomers were also telling Einstein that the universe was static, so he wanted to guarantee the theory would agree with observations on stars and galaxies. (Later Einstein called his cosmological constant 'the biggest mistake of his life'.) But in 1922, the Russian Alexander Friedman derived a set of equations that showed that it *was* possible for the universe to expand, and that the universe was not static after all.

Friedman was a Russian cosmologist and mathematician who lived much of his life in Leningrad. He fought in the First World War as a bomber and later lived through the Russian Revolution of 1917. He obtained his degree in St Petersburg State University (1910) and eventually became a lecturer in St Petersburg State College of Mines. He became a professor at Perm State University in 1918.

In addition to general relativity, Friedman's interests included hydrodynamics and meteorology. In June 1925 he was appointed as the director of Main Geophysical Observatory in Leningrad. Shortly after, in July 1925, he participated in a record-setting balloon flight (recording an elevation of 24,000 feet (7,315 metres)). Friedman died on 16 September 1925, at the age of only thirty-seven, from typhoid fever contracted during a vacation in Crimea. Since the distribution of mass in the universe determines the curvature of space in Einstein's theory, Friedman made the simplifying assumption that the universe was uniformly filled with a thin soup of matter. Modern measurements have shown this assumption of uniformity to be quite reasonable in spite of the formation of stars and galaxies. Under this assumption, Friedman found that general relativity predicted the universe to be unstable and the slightest perturbation would cause it to expand or contract.

Friedmann did not include lambda in the equations he solved, as he considered the cosmological constant an arbitrary addition that was not necessary. He subsequently predicted an expanding universe for the particular mass density he initially chose. This Einstein did not like. But the Russian's work also unveiled a very interesting characteristic of the equations. He showed that for different mass densities, the rate of expansion or contraction and the size of the universe and its ultimate fate would be different. Since mass produces the curvature of space, the amount of curvature depends on the amount of mass present in the universe. Friedman's conclusions can be summarized by considering three different values for the mass density of the universe expressed as a ratio of a critical value, Ω, the Greek letter omega.

If mass density is greater than the critical value, $\Omega > 1$

In this first case, the expansion rate is slow enough and the mass great enough for gravity to stop the expansion and reverse its direction; a so-called 'big crunch' would eventually result with all the matter in the universe pulled back to a single point.

If mass density is less than the critical value, $\Omega < 1$

In this second case the universe expands much more rapidly. Gravity can't stop it but does slow the rate of expansion somewhat.

Mass density is equal to the critical value, $\Omega = 1$

Finally, in this case the universe expands just fast enough not to collapse. The speed at which the galaxies recede from each other gradually decreases, and galaxies move apart from each other for all time. This characteristic defines the 'critical' value.

Friedman's solution resulted in the surprising prediction that the universe can expand *or* contract. Moreover, he posited that the rate at which this happened would depend on the total amount of matter contained in the original universe.

More so-called 'universes' would soon be derived from the same equations. These would be the result of work by the

British astronomer Arthur Eddington, the Swiss mathematician Hermann Weyl, a Hungarian-German physicist named Cornelius Lanczos and finally the Belgian astrophysicist Georges Lemaître. Later Americans Howard Robertson and Richard Tolman would also add yet another version of the universe based on Einstein's same field equations.

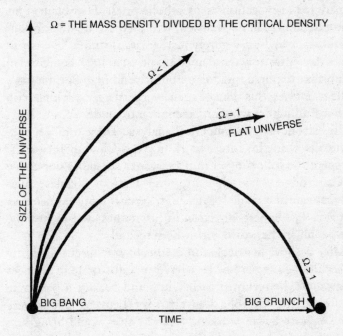

Abbé Georges Lemaître

The year 1927 saw the arrival of an unusual new protagonist in the story of astronomy. Abbé Georges Lemaître, a Catholic priest from a small university in Belgium, solved Einstein's equations in a manner similar to, but completely independent from, Friedman five years earlier. Lemaître's contribution is vital because it added some essential and much-needed physics to the mix. Lemaître also became a pioneer in the application of the field equations to the newly-respectable field of cosmology.

After a classical education at a Jesuit secondary school, Lemaître began studying civil engineering at the Catholic University of Leuven at the age of seventeen. In 1914 he interrupted his studies to serve as an artillery officer in the Belgian army and served for the duration of the First World War. After the war, he studied physics and mathematics, and began to prepare for his vocation as a Catholic priest. He obtained his doctorate in 1920 in Belgium and wrote a thesis on functions of several real variables, a purely mathematical study.

Ten days after he was ordained as a priest in 1923, he arrived in Cambridge to pursue graduate studies, residing at St Edmund's College. During this time he spent a crucial year working with Arthur Eddington, who initiated him into modern cosmology, stellar astronomy and numerical analysis. These topics would eventually shape his future work in physics and mathematics. He spent the following year at Harvard College Observatory in Cambridge, Massachusetts, with Harlow Shapley, who had just gained a name for his controversial work on nebulae. Lemaître had superb direction in his studies, and picked the best people in the world with whom to study.

After a short matriculation at the Massachusetts Institute of Technology, where he registered in a doctorate in sciences programme, he returned to Belgium and became a part-time lecturer at the Catholic University of Leuven. It was here that Lemaître began work on a report that would bring him international fame. In this report, he presented his idea of an expanding universe similar to Friedman's, but avoiding the problems of Einstein's finite, closed and static universe. Published in 1927 in *Annales de la Société Scientifique de Bruxelles* (Annals of the Scientific Society of Brussels), the report was titled 'A Homogeneous Universe of Constant Mass and Growing Radius Accounting for the Radial Velocity of Extragalactic Nebulae'.

In this calculation, he derived the relation for an expanding universe to be between the speed of a galaxy receding from an observer and its distance from the observer. Lemaître also provided the first observational estimate of the slope of the speed-distance curve that later became known as Hubble's law when the American astronomer Edwin Hubble reported his initial observations on galaxies in 1929. These two important properties of the universe were proposed two years before the measurements that would begin a new era in astrophysical cosmology. As a postscript to his paper, Lemaître also made reference to the fact that, 'we still have to explain the cause of the expansion of the universe'. These were revolutionary, prescient ideas that turned out to be mostly correct. Unfortunately, the paper had little impact internationally because the journal in which it was published was not widely read by astronomers outside of Belgium. Einstein, who was then in nearby Germany attending conferences and meetings sponsored by the Solvay Institute in Belgium, could not argue with the mathematics of Lemaître's theory. But he did refuse to accept the idea of an expanding universe. Lemaître recalled Einstein commenting in French, *Vos calculs sont corrects, mais votre physique est abominable* (your maths is correct, but your physics is abominable).

Things changed two years later, however, when Hubble published his systematic observations of galaxies and confirmed the expansion of the universe. The Royal Astronomical Society immediately gathered to consider the seeming contradiction between visual observation and the theory of relativity that Hubble's findings threw up. When Lemaître read about these proceedings, he immediately sent Eddington a copy of his 1927 paper. The British astronomer realized that Lemaître had bridged the gap between observation and theory and immediately shared Lemaître's paper with the Dutch cosmologist de Sitter, who soon after wrote to Harlow Shapley at Harvard:

I have found the true solution, or at least a possible solution, which must be somewhere near the truth, in a paper ... by Lemaître ... which had escaped my notice at the time.

Einstein soon confirmed that Lemaître's work 'fits well into the general theory of relativity'. Later de Sitter publicly praised Lemaître's 'brilliant discovery, the "expanding universe"'. At Eddington's suggestion, an English translation of Lemaître's paper was published in the Royal Astronomical Society's journal *Monthly Notices* of March 1931.

In the same year, the Belgian cosmologist published another important paper in the prestigious British journal, *Nature*. In it he expounded his concept of the primordial atom or *cosmic egg*. He proposed that the universe had expanded from an initial point, which he called the primeval atom. He also described his theory as 'the cosmic egg exploding at the moment of the creation'. (This would later become known as the Big Bang theory, which exploded sometime after the beginning of time into the expanding space to produce the matter which makes up the universe.)

Lemaître's ideas produced yet more questions, many of which forced physics and astronomy together: What was that primordial atom like? Why would it explode? He pursued the topic for some time, going so far as to suggest that there ought to be some form of background radiation in the universe, left over from the initial explosion of that primordial atom. (This background radiation was not observed until 1965.)

It is for this cosmic egg description that Lemaître has been called, 'The Father of the Big Bang', though the details of the early universe have been revised considerably from then up to the present day. A new concept, starting at time = 0 with a hot, dense, primordial soup phase and an accelerated expansion called inflation, is the currently accepted standard model of the early universe replacing Lemaître's first picture.

In January 1933, both Lemaître and Einstein travelled to California for a series of seminars that Edwin Hubble also attended. After the Belgian detailed his theory, Einstein stood up, applauded, and said, 'This is the most beautiful and satisfactory explanation of creation to which I have ever listened.' An article about Lemaître appeared on 19 February 1933 in the

Eddington meets Einstein at Leiden Observatory, September 1923. Front row: Arthur Eddington and Hendrik Lorentz; Back row: Albert Einstein, Paul Ehrenhest and Willem de Sitter. All of these men made important contributions to General Relativity.

New York Times Magazine, featuring a large photo of Einstein and Lemaître standing side by side. The caption read: 'They have a profound respect and admiration for each other.' In the same year, when he published a more detailed version of his theory in the *Annals of the Scientific Society of Brussels*, Lemaître would achieve his greatest glory. Newspapers around the world called him a famous Belgian scientist and described him as the leader of the new cosmological physics.

Lemaître received many awards, including the very first Eddington Medal awarded by the Royal Astronomical Society. He died on 20 June 1966, shortly after learning of the discovery of cosmic microwave background radiation, which provided evidence that his intuitions about the birth of the universe were mostly correct.

11

THE VIEW FROM MOUNT WILSON

Observations of the Universe 1900–33
At the end of the nineteenth century, it became well known amongst astronomers that the mountains of California, with their stable and transparent atmosphere, were ideal for celestial observations. Steady air in combination with dark skies resulted in what astronomers call 'clear views', with a high percentage of 300 clear nights per year. The first major effort to build a large-scale telescope here came after a bequest of over US$700,000 from the passionate astronomy enthusiast and piano manufacturer James Lick, who wished to have an observatory named in his honour. Constructed on Mount Hamilton in the Diablo Range just east of San Jose, the observatory was founded in 1888 and has been part of the University of California ever since. So important was Lick's bequest to the realization of his dream, that when Lick died, his body was buried under the future site of the telescope with a brass tablet bearing the inscription, 'Here lies the body of James Lick.'

As part of the University of California, the observatory became the first permanently occupied mountaintop observatory in the world. The first telescope built at Lick was a refractor with a glass lens diameter of 35.8 inches (0.91 metres), slightly smaller than the largest practical functioning refractor at the Yerkes Observatory in Chicago, which had a 40-inch (1.02-metre) diameter lens. The commissioning of a new 91–inch (2.3-metre) reflecting telescope at the Lick in 1900, twelve years after its founding, was indeed an auspicious beginning of the new age. Here, Director James Keeler and his colleagues were able to obtain spectacular images of spiral nebulae, including the famous image of M51, the Whirlpool Galaxy. Details of the structure of this galaxy could be seen quite clearly at this early date, as were images of more faint spiral nebulae. The possibility that these nebulae, or clouds, were similar to the Andromeda galaxy and might lie a very great distance from our solar system became the first point of enquiry for the great telescope.

Telescopes had been essential to astronomers since Galileo built the first model in 1610. These simple tubes – which used to be hand-held to make observations – over the next four hundred years had come a long way. Telescopes now tend to be huge devices, constructed at a cost of millions of dollars. The most important feature of the telescopes used for the critical astronomical observations made in the first few decades of the twentieth century was the aperture, or the size of the opening in the telescope for the collection of light. Since the amount of light energy that is concentrated into the image plane in today's large telescopes determines how much time is required to record the image electronically or photographically, aperture becomes a very important factor. The telescope must also follow the object it is tracking in the sky at the same rate as the rotation of the earth, even during the daylight hours for very long exposures, which in some cases, lasts days. Of course the aperture itself is closed during the daylight period. The image of a star can be as small as a point, and the total amount of light energy to be analysed depends only on the area of the mirror.

At the beginning of the twentieth century it was obvious to those with an interest in astronomy that sensitive light gathering instruments would increasingly be needed to provide new information on the makeup and structure of the universe – especially as astronomers searched deeper and deeper into the darkness of space for information. There is hardly a more exciting example of this fact than in the history of the Mount Wilson Observatory, which was the second major site constructed in the California mountains, not far from the Lick Observatory on Mount Hamilton.

George Ellery Hale was born into a wealthy Chicago family in 1868 and from an early age was obsessed by science and particularly, astronomy. His interest led him to construct his own observatory at age twenty, at the Hale family home. There, in 1888, he acquired a professional long-focus refractor and spectroscopic apparatus that were as good as the equipment of most colleges at that time. When he graduated from the Massachusetts Institute of Technology with a bachelor's degree in physics in 1890, Hale presented in his senior thesis a design for a *spectroheliograph*, an instrument for photographing the Sun in a very narrow range of visible wavelengths, that is, in nearly monochromatic light. He would use this device to measure the magnetic fields on the surface of the Sun, exploring sunspots.

Hale's work and his observatory came to the attention of William Rainey Harper, the first president of the new University of Chicago, which was funded by millionaire John D. Rockefeller. Harper attracted Hale and his observatory to the university in 1892, a truly unique start for a young astronomer. Through Harper, Hale would soon come into contact with wealthy philanthropists like Rockefeller. This would be important in the future career of the ambitious young man. Wasting no time, Harper and Hale secured support from the transportation magnate Charles T. Yerkes in October of that same year to build a great observatory at Chicago. The new facility, named after Yerkes, would have space and 'laboratories for optical, chemical and spectroscopic' work – not just for observing. The telescope would be a 40-inch (1.02-metre) refractor, just a few inches larger than the first refractor at

Lick. From a young age Hale was aware of the importance of size when building a telescope, This would be the first of four of the 'largest [telescopes] in the world' that were built during his fifty-year career.

Ten years later, Hale was a full professor at Chicago and director of the important Yerkes Observatory on the campus. But this did not satiate Hale's desire to further his knowledge. He learned that Andrew Carnegie had established the Carnegie Institution of Washington to support original research in all scientific fields, and immediately saw the possibilities for furthering his own research in the solar spectrum and the magnetic fields of sunspots. Hale felt strongly that new telescopes should be designed specifically to suit the problems they are meant to investigate, rather than the other way around. As powerful as telescopes such as the Lick 36-inch (.92 metre) and the Yerkes 40-inch (1.02-metre) might be, they were not well suited to the study of the Sun, which was Hale's main special interest. More importantly to the study of the structure of the universe, Hale also felt the future of stellar spectroscopy, that is photographing the spectrum of light emitted by stars, would best be served by large mirrored reflectors, not bulky glass refractors. Refractors had reached their practical size limit of 40 inches (1.02 metres), but more light was needed if the stars' spectra were to be examined in greater detail.

In 1896 Hale's father made another of his generous contributions to his son's career: a 60-inch (1.52-metre) optical glass disc to be used as a substrate for the mirror of a giant reflecting telescope. Had Hale's father not died soon after, he may have provided funds for the mounting of the telescope, but instead, the disc lay in the Yerkes Observatory basement while Hale searched for a donor to complete the project. It was thus that Hale managed to get himself appointed to the Advisory Committee on Astronomy of the new Carnegie Institution, joining Edward Pickering from Harvard, as well as other experts. This seemed a good opportunity to influence how the Carnegie money could be spent. However, Hale was alone in arguing that new methods were necessary in order to advance an understanding of the Sun and stars, a field which later became known as astrophysics.

Consequently, a group from the advisory committee was formed to research the establishment of an observatory. They were sent to investigate several sites in southwestern US, including Flagstaff in Arizona (home of the future Lowell Observatory), as well as Mount Palomar and Mount Wilson in California. Mount Wilson received the highest recommendation and Hale decided to visit there himself. On 25 June 1903 Hale journeyed up the mountain with W.W. Campbell, Director of Lick Observatory. Low clouds and fog covered Pasadena that morning as Hale, depressed by the gloomy skies and doubting that Wilson's Peak would provide a satisfactory site for his solar observatory, started up the trail. All of his misgivings were replaced with delight, however, when half-way up the mountain the weather burst into blue sky and brilliant sunshine. With the 9-inch refractor telescope that had been set up for him on the top of Mount Wilson, Hale proceeded to make his first solar observations.

Ecstatic, Hale immediately decided he must build his observatory on Mount Wilson and soon after his return to Chicago, applied to the Carnegie Institution for a grant to build a great solar observatory. Without waiting for a response, he moved his family to Pasadena six months later. Hale was sure that his application for a solar observatory would have a good chance to win some Carnegie funding, and on 13 June 1904, he signed a 99-year lease with the Mount Wilson Toll Road Company for access to the mountain. By the time the Carnegie Board of Trustees met in December of that same year, Hale had already spent US$27,000 of his own money to finance his dream. His determination was rewarded when the Carnegie Institution of Washington agreed to fund the Mount Wilson Solar Observatory and appointed Hale the founding Director. One year later in 1905, (whilst Einstein was publishing his first paper on relativity), the observatory became operational.

Some thought Hale was a reckless gambler, but others saw

him as a genius and visionary who simply could not accept the possibility of failure. As the first observations were being made on top of the soon to be legendary mountain, Hale was already making plans to bring the 60-inch (1.52-metre) disc out of storage at Yerkes to Mount Wilson and build the world's greatest telescope. On 8 December 1908, first light was transmitted through the 60-inch (1.52-metre) telescope. Only five years after Hale's first visit to Mount Wilson, the first photographs of the heavens were taken with the mirror made from the glass disc which Hale's father had purchased for him in 1896.

By the age of thirty-six, George Ellery Hale had moved away from active research in astronomy to become an entrepreneur in the development of the world's biggest telescopes. Ever successful at inspiring wealthy businessmen to support his work, Hale had moved on to another benefactor: the elderly Los Angeles businessman John D. Hooker. Hooker financed Hale's next project, the 100-inch (2.54-metre) telescope that transformed observational cosmology during the years from its commissioning in 1918 until 1948. The glass for the 100-inch (2.54-metre) mirror, purchased from Saint-Gobain glass works in France, arrived with much fanfare in Pasadena on 7 December 1908, a day before the 60-inch (1.52-metre) scope became operational.

Hale's approach was to remain one step ahead of everyone else, and with one of the wealthiest men in the world as a sponsor, they were a formidable team. As one large telescope was completed, plans would be in development for an even larger one. One such telescope would have been enough to assure Hale a place in history, but he planned one scientific breakthrough after another on Mount Wilson. Four years after the observatory was begun, four great telescopes, including the world's largest solar and stellar instruments, would reside there. Many fundamental problems in astronomy – the nature of sunspots, the temperature and composition of the stars, even the structure of the universe – would eventually be addressed on this mountain.

Carnegie, the generous sponsor of the entire observatory,

visited Mount Wilson in March 1910 and saw the 60-inch (1.52-metre) telescope in operation and the Hooker 100-inch (2.54-metre) under construction. The proud philanthropist posed for a photograph with Hale, announcing that the Carnegie Institution had became a permanent part of Mount Wilson. When the 100-inch (2.54-metre) reflector finally became operational in 1918, Hale had for a third time built the largest telescope in the world.

Throughout the 1920s, Hale wrote a series of popular articles about the possibilities of large telescopes, waxing romantically about the many compelling rationales underpinning astronomy's insatiable need for light-gathering power. In 1928 he attracted some US$6 million from yet another wealthy philanthropic source – the Rockefeller Foundation's International Education Board – for the construction of a 200-inch (5.08-metre) reflector on another mountain in California, Palomar. This was a major coup at a time when overall support for science in the US was waning. The next two decades would uncover many technical and social obstacles to the completion of the telescope. A serious setback occurred when Hale died in 1938 and construction was halted altogether during the Second World War (1939–45). Nevertheless, in 1948 the new telescope, suitably named after Hale, saw first light at the Palomar Observatory. It remained the largest telescope in the world until 1976.

Hale's legacy with regard to large telescopes is indeed impressive. But he was also a prolific organizer, and helped to create several first-class astronomical institutions, formed astronomical societies and edited journals. He also played a central role in developing CalTech, the California Institute of Technology into a leading research university. Hale's role in the discovery of the universe was not just about building telescopes, raising money and sitting in boardrooms. Though he was a solar specialist in that he mainly studied the surface of our closest star, he knew that an understanding of the structure of the universe had to do with knowledge of the characteristics of the other billions and billions of stars in the sky. If his observatory was going to make a major contribution to this search, he would not only need the best equipment to perform these studies, but the best astronomers to work with him on top of Mount Wilson. He was an indefatigable networker and knew everyone who was important in astronomy in the early part of the twentieth century. And of course, everyone wanted to know him.

Hale took great care in making sure that the observatory was adequately staffed. He made a point of hiring astronomers and physicists who could both pursue independent research and contribute to the MWO (Mount Wilson Observatory) community. He also engaged a group of over a dozen 'human computers', almost always women, whose job it was to analyse the observational data provided by the astronomers. With his great authority, Hale was able to convince the Carnegie purse holders to support special research associates, visiting scientists funded separately by the institution, whose work would invariably be carried out in connection with the MWO.

Some of the more important of these associates include Albert Michelson and Henry Norris Russell, each internationally-renowned in the world of astronomy. In the 1920s, they were both involved in unique projects to further man's general understanding of the stars, something close to Hale's heart. Hale dropped the term 'Solar' from the title of the observatory so that its new focus increasingly became that of astrophysics and cosmology, i.e. everything in the universe. Though the main activity at MWO was stellar spectroscopy, Michelson's attempts to measure the size of stars and Russell's goal to develop a theory of stellar evolution, both fit well into Hale's overall plan.

The 'Research Associate' concept clearly paid off. In December 1920, Michelson and MWO staff member Francis Pease made the first determination of a stellar diameter, measuring the red giant Betelgeuse in the constellation Orion using an astronomical interferometer and the 100-inch [2.54-metre] Hooker telescope. The interferometer was a device that used the interference patterns produced by superimposing light beams to increase the sensitivity of a measurement. Michelson, the master of interferometer techniques, had used the method in 1887 to show that the infamous 'luminiferous ether' could not be detected in the famous Michelson–Morley experiment. Michelson and Pease built the massive interferometer contraption on top of the Hooker. They found, to the amazement of the world, that Betelgeuse was so large that if it was positioned at the centre of our solar system, its surface would extend out to the orbits of Mars and Jupiter, wholly engulfing the orbits of Mercury, Venus, Earth and Mars.

Russell, unlike Michelson, was more than an occasional visitor to Mount Wilson. Starting in 1921, Russell spent a month or more on the mountain almost every spring for thirty years. He pushed the spectroscopists to produce more and more data in pursuit of his lifelong goal: To move from measurements of the properties of stars to an understanding of the physics of stars and a theory of stellar evolution.

Classifying the Stars: H.N. Russell

Science historian David H. DeVorkin has characterized Henry Norris Russell as 'America's most brilliant and powerful astronomer of the first half of the twentieth century'. Russell obtained his PhD from Princeton in 1900 and immediately left for the University of Cambridge, England. There he concentrated on theory, namely the mathematics of orbit theory and celestial dynamics, and observations, particularly photographic studies for determining the distances to stars using the parallax method.

When he returned to Princeton in 1905 as a junior faculty member, Russell believed that the future of astronomy was not in mere data gathering but in focused research in which theory and observation would work together on a specific problem. This later became the norm in twentieth century astronomical research. He was fortuitous in that at Princeton, research was not instrument-based nor was it solely defined by the interests of the observatory director. Russell was therefore free to search out new and exciting problems and to apply his considerable mathematical talents to their solution.

Russell spent nearly his entire professional life at Princeton and rose quickly through the ranks, securing a professorship in 1911 and director of the observatory a year later. Although he maintained this position until his retirement some thirty-six years later, his chief activity was always research, avoiding administration and teaching whenever possible. He did little student supervision, but his prize graduate student Harlow Shapley went on to work with Hale at MWO during 1914–17 and carried out an historic project on the size and configuration of the Milky Way galaxy.

Until 1920 Russell's research interests ranged widely within planetary and stellar astronomy and astrophysics. He worked mainly on binary stars and what occurs when they appear to move in front of each other as they orbit about their common centre of gravity. He developed statistical methods for estimating the distances, motions and masses of groups of these binary stars. Following up on his post-doctoral work in England, Russell applied his study of binary stars to investigate what they could reveal about the lives and evolution of stars and stellar systems. He used parallax distance measurements to determine the absolute brightness of these stars and then compared their brightness to their colours, or spectra. He found that among the majority of the stars in the sky, blue stars are intrinsically brighter than yellow stars and yellow stars are brighter than red stars. He found that a few exceptionally bright yellow and red giant stars, like Betelgeuse measured by Michelson, did not follow this relationship at all.

By collecting many years' worth of data from his many colleagues at US observatories, he was able to plot brightness against colour (spectra), in a diagram illustrating the definite relationship between a star's true brightness and its spectrum. He announced these results in 1913 by way of a simple diagram, which was called the 'Russell Diagram' by American astronomers. However, the Danish astronomer Ejnar Hertzsprung had obtained the same result several years earlier, the diagram has now become known as the Hertzsprung–Russell (HR) diagram, and is now an important feature of modern astronomy.

Today an astronomer can measure the light from a star to determine its spectral type. If the star lies on the main sequence, the astronomer determines its absolute magnitude (the true brightness) from the standard HR diagram. By then measuring the star's apparent magnitude, the distance to the star can be calculated using a simple mathematical relationship called the inverse square law. Thus by simply measuring the colour of light from a star, the distance to the star can be found using the HR diagram. Most stars lie in a band from the upper left to the lower right. In this band, blue stars are brighter and red stars are fainter, with white and yellow stars falling somewhere in between. This band is called the 'main sequence'. Although Hertzsprung and Russell didn't know it at the time, stars on the main sequence are in their youth and middle age, which is also when they burn hydrogen in their cores. As a star begins to burn helium and heavier elements in the core, it quickly evolves off the main sequence into other types of stars like giants, super giants and eventually white dwarfs.

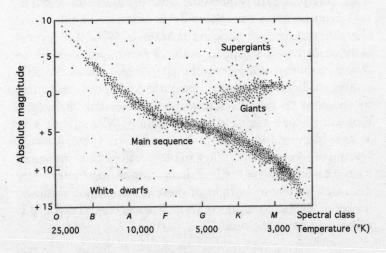

HR diagram showing the sequence of the stars.

Like his friend George Ellery Hale, Russell was hyperactive and nervous, filled with ideas and subject to occasional breakdowns. While Hale developed his skills as a superb fundraiser, organizer and administrator, Russell saw himself strictly as a scientist and teacher. He depended on others to develop and manage the observatories and then provide him with data for his research. Clearly, he had to be a first-class theorist to be so presumptuous. Yet Russell also made himself accessible to his widening circle of astronomical colleagues, so he achieved a balance in his approach. He soon found that observational astronomers were happy to offer their hard-won data to someone so talented and prestigious. He soon became Hale's chief theorist and received full cooperation from the astronomers at MWO.

With the energetic and brilliant leadership of the American astronomical establishment – led by figures such as Hale and Russell – the attack on understanding the characteristics of stars proceeded at a pace. Cosmologists soon began to want to know the size and temperature of stars, their composition and colour and their evolution. Two key questions about the stars were particularly pressing: their position and distribution. How far away were the millions of stars from the Earth? Were they moving or static in the heavens? What was desperately needed were signposts in the heavens – markers which would enable astronomers and their evermore powerful telescopes to determine a star's position and speed. Advances in telescope construction anticipated these new challenges, especially those taking place at Mount Wilson, but also at the Harvard, Lowell, Yerkes and Lick Observatories. During the nineteenth century so few stars had been measured by the direct parallax method, indicating that the interstellar distances were vast. In light of these disappointing findings, however, Einstein reminded his colleagues that, 'The Lord is subtle, but not malicious.'

Signposts in the Heavens

Anyone who has ever practised their long-distance golf shots off the tee of a driving range will be familiar with the signposts that are distributed to measure distances: These '100 metres', '150 metres', '200 metres' markers indicate the distance from the tee. When a golf ball lands, the closest sign will enable the player to estimate immediately the distance of the drive. Astronomers began asking themselves whether such a marker could ever to be found in the cosmos to estimate the distance to stars and nebulae. The answer would turn out to be yes. An unusual type of star called a Cepheid variable was discovered in 1912, by Henrietta Leavitt, a determined and thorough investigator at the Harvard College Observatory.

The Cepheids take their name from Delta Cepheid, a star in the constellation Cepheus, which was first discovered in 1784 by the young Englishman John Goodricke. Because Delta Cepheid is the prototype, many books and articles state that it was the first Cepheid to be located. But this honour belongs to another bright star, Eta Aquilae. Eta Aquilae's variability was spotted more than a month earlier by Edward Pigott, a wealthy young British amateur in York. Astronomers of Pigott's time knew of only five variable stars: Mira, P Cygni, Chi Cygni, R Hydrae and Algol. But Pigott rightly suspected there were many more. In 1781 he began a campaign both to monitor the known variable stars and to hunt for new ones.

Collaborating with Pigott in York was John Goodricke, who was only seventeen years old and was born deaf and mute. In eighteenth-century England, this disability was associated with low intellectual ability. But Goodricke's parents refused to accept this view and set him on a vigorous academic course. He developed a passion for science, especially astronomy. Pigott became Goodricke's friend and mentor and taught him how to study the sky. Sometimes the two observed together from Pigott's well-equipped backyard observatory three blocks from Goodricke's home in York. Other times they observed separately and compared notes.

On the night of 10 September 1784, Pigott discovered that Eta Aquilae changed brightness. He noticed this while monitoring another star, Theta Serpentis. Comparing the two, Pigott noticed that the former looked fainter than it did the year before and soon confirmed the star's variability. In the process of so doing, he also discovered what astronomers now recognize as trademark Cepheid behaviour: the star rises fast to maximum brightness, then falls slowly to minimum in a very regular cycle.

The very same night that Pigott discovered Eta Aquilae's variability, Goodricke noticed an alteration in the light of what would become another prototype variable, Beta Lyrae. A month later, on 20 October, Goodricke discovered the variability of Delta Cephei when he noted that one of the stars of Cepheus did not look quite right. Three nights later he wrote that he was 'almost convinced' that the star doing the changing was Delta.

That winter Goodricke continued to observe the star which, unlike Pigott's Eta Aquilae, is circumpolar from England (it never sets below the horizon because of its proximity to the north celestial pole). It could thus be followed all year long for the entire night on every night of the year. He monitored it night after night and soon found a distinct variation in brightness with a period close to the modern accepted one of 5.37 days. Sadly, Goodricke died in 1786 at the age of only twenty-one years old, just as he was about to embark on an important scientific career. Pigott wrote of his death that, 'I had the misfortune to lose the best of friends . . . which took away all the pleasure I ever had in astronomical pursuits.'

Delta Cepheid is now the prototype of the Cepheid variables, and is the closest star of this type to the Sun. These stars are yellow super giants, and their fluctuations in brightness results from an actual physical pulsation with corresponding changes in surface temperature and size. This star formed part of the original 1908 study, in which Henrietta Leavitt first discovered that the periods of luminosity were related to their absolute luminosity. This has proved to be a most important distance-measuring tool.

Henrietta Leavitt began her education in 1886, at Oberlin College in Ohio, attending for two years the school that was noteworthy for its admission (from as early as 1834) of African-Americans and women. She then transferred to the Society for the Collegiate Instruction of Women, which had become Radcliffe College by the time that she graduated in 1892. Following an interest in astronomy that she discovered in her senior year, she became a volunteer assistant at the Harvard Observatory in 1895, and joined the legendary Annie Cannon as a member of Edward Pickering's team of assistants. In 1902 she received a permanent staff appointment and from the outset was employed in Pickering's great project to determining the brightness of all measurable stars. Pickering, who was then the world's expert in photographic photometry (determining the brightness of a star from its photographic image), did not take Leavitt seriously because she was a woman. He assigned her the tedious task of cataloguing 'variable' stars, whose brightness appears to ebb and flow in predictable patterns.

Taking her assignment beyond its original brief, Leavitt decided to try to determine the relationship between the brightness of a variable star, referred to as a Cepheid, and the period of its variation. To do this, she would need to study many different Cepheids, measuring the brightness and period for each one. She immediately encountered a problem: The observed brightness, or apparent brightness, of a star depends strongly on the distance between the star and the observer. In her proposed study, she would therefore not be able to ignore the distance variation factor unless all the Cepheids she measured were, approximately at least, the same distance away.

It was at this point that she had a stroke of genius. She considered the well-known Magellanic clouds of stars. Since these were known to be small stellar systems, it would be reasonable to assume that the stars in these clouds are all about the same distance from the Earth. She deduced that the difference in the *apparent* brightness would therefore correspond directly to differences in the *inherent* brightness or 'luminosity' of the star and not due to any distance factor. She was able to make this deduction simply because all the stars in the Magellanic cloud are about the same distance away from the Earth.

The Magellanic clouds are not even visible from the northern hemisphere, but hundreds of photographic plates of stars in those clouds had been obtained between 1893 and 1906 at Harvard College's observatory in Peru, including a catalogue of 1,777 variable stars. Among these, sixteen stars appeared in a sufficient number of plates for their periods to be determined. When tabulated in order of increasing brightness, or luminosity, a pattern emerged. By 1908, Leavitt observed the results with typical modesty and understatement when she wrote: 'It is worthy of note that the brighter variables have the longer period.' Four years later in 1912, Leavitt produced more data on the period vs. luminosity (P-L) relation and had by this time also managed to obtain data on twenty-five Cepheid variables in the Small Magellanic Cloud (SMC). She wrote:

A remarkable relation between the brightness of these variables and the length of their periods will be noticed. In 1908, attention was called to the fact that the brighter variables have the longer periods, but at that time it was felt that the number was too small to warrant the drawing of general conclusions. The periods of 8 additional variables, which have been determined since that time, however, conform to the same law.

She further added:

... Since the variables are probably at nearly the same distance from the Earth, [all are in the same small Magellanic cloud], their periods are apparently associated with their actual emission of light, [or luminosity] as determined by their mass, density, and surface brightness.

She produced graphs of the P-L relation and wrote:

A straight line can readily be drawn among each of the two
series of points corresponding to maxima and minima, thus
showing that there is a simple relation between the brightness of
the variables and their periods.

As this period-luminosity curve applies to all Cepheid stars,
it would *eventually allow astronomers to measure the distance
from the Earth to any visible Cepheid star in the universe*.

One major problem remained. Measuring two different
Cepheids gives only the relative distances from the observer of
the two stars. How can the Cepheids be calibrated to yield the
actual distances? To do this, the actual distance to at least one
star would need to be independently and reliably measured.
Ideally observers would source a Cepheid in close proximity
to another star whose distance could be determined by the
parallax method, for example, which is accurate and reliable.

This was eventually accomplished by the Danish astronomer
Ejnar Hertzsprung, who shared the credit for the discovery of
the Stellar Sequence Diagram with H.N. Russell. Hertzsprung
realized that if the P-L relation could be calibrated, then the
stellar luminosity of this group of variable stars might be determined directly from their periods. He was able to estimate the
distance to the Small Magellanic Cloud (SMC) using a statistical parallax method and determined the true distance to a
single Cepheid, thus calibrating Leavitt's relation. It was then a
straightforward matter to obtain the distances for all Cepheids
using the inverse square law.

The Yardstick of the Universe

Cepheid variables are the most important type of variable
because their periods of variability are related to their absolute
luminosity or brightness. This makes them invaluable as tools
to measure astronomical distances. The periods are very regular and range from 1 to 100 days. These stars have proved to be

the most valuable methods for distance determination because their period of variability is related to their absolute brightness by a simple period-brightness relationship. The stars can then be calibrated as standard candles for distance calculations since measuring their apparent brightness allows a straightforward calculation of their distance.

By simply measuring how bright the star appears, one can figure out how far away it must be as the brightness obeys the simple inverse square law. For example, if the distance to a star is doubled, its brightness reduces by one-fourth; if the distance is tripled, the star will be one-ninth as bright. Today Cepheid variables can be seen and measured out to a distance of about 20 million light years, compared to a maximum distance of about 65 light years for Earth-based parallax measurements. Period-brightness relation is one of the backbones of modern cosmology and is used to calculate the distances to other galaxies over vast dimensions of space.

In the course of her work, Leavitt also discovered four novae and some 2,400 variable stars (about half of the total known at the time). She died of cancer in 1921 before her work on a new photographic brightness scale was completed. Her death was viewed as a 'catastrophe' by many of her colleagues. However, her important contribution to scientific advancement was internationally acknowledged in 1925 when the Swedish Academy of Sciences nominated her for the Nobel Prize. This is ironic given that the Nobel is never given posthumously. She was so relatively unknown that the Swedes did not even know she had been dead for four years.

Percivial Lowell and Vesto Slipher
A second type of 'signpost in the heavens' required by cosmologists was some kind of automatic velocity detector which would enable long-range telescopes to photograph stars. These photographs could then determine if a star is moving or stationary. Impossible? Well, it turned out to be no more impossible than finding the Cepheids. Consider the modern gadget used

by traffic police to photograph a speeding automobile. Many will recall the unsettling experience of driving through a speed zone, and realizing that a flash has just gone off. This flash belongs to a special device that photographs automobiles and their respective licence plates, simultaneously linking these photos with a photo displaying the speed of travel. This technology allows traffic police to ascertain whether the speed limit has been exceeded, and to issue fines accordingly.

This second type of 'signpost in the heavens', a photograph which could instantaneously measure the speed of a star, was actually discovered by a young astronomer at the Lowell Observatory in Arizona in 1912. (This was the same year Henrietta Leavitt announced her discovery of the Cepheids.) In 1894, Percival Lowell had established an observatory in Flagstaff, Arizona, to study the planets and to search for life in the universe. Lowell thought that the spiral nebulae being reported by astronomers at that time might be other solar systems in the process of formation. He therefore asked one of the observatory's youngest astronomers, Vesto M. Slipher, to photograph the light from the diffuse spiral nebula to see if their spectra suggested chemical compositions like those expected for newly forming planets.

The Lowell Observatory's major instrument was a 24-inch (.6-metre) refracting telescope, which was not at all suited to observations of faint spiral nebulae. Despite this, Slipher discovered that some of the diffuse nebula had dark line spectra. With the technology available at that time, he was able to take photos with exposures lasting 20–40 hours. His first spectrum in 1912 was of the Andromeda Nebula, M31, which was not known to be a galaxy at that time. Its spectra did not suggest new planets but did display a shift of the absorption lines. This shift indicated a motion of that object toward the earth at about 300 miles-per-second (480 kilometres-per-second) using the Doppler formula. This would later be called 'blue-shifted' because the lines displace toward the blue end of the spectrum.

In 1842 Austrian mathematician and physicist Christian

Doppler pointed out that if a light source is approaching an observer, the light waves will be crowded more closely together or, if the source is moving away, will be spread out. The principle became known as the Doppler effect. Naturally, if the source is stationary with respect to the observer, the successive wave crests will be spread out evenly in all directions like the ripples from a splash in a pond. However, a moving source results in a decrease or increase of the length of light waves depending on the direction of the motion in relation to the observer. For astronomers looking at the distant stars this motion is called the radial velocity, the motion along the radius away from or towards the Sun. This effect is the same for sound waves: Standing by the highway the characteristic increase and decrease in the pitch of the siren as an ambulance passes by is clearly audible. The pitch is higher than normal when a siren approaches and lowers when in recedes.

Consider the spectrum of light coming from a star. The colours of its spectrum range from blue (short wavelength) to red (long wavelength). As the above description details, when a star (or galaxy) moves towards an observer on Earth, it will shorten its wavelength and shift the spectrum slightly towards the blue. Movement away from the earth will have the opposite effect, and shift the spectrum towards red. Unfortunately, a small shift in the continuous spectrum of the starlight – unless the star's speed is tens of thousands of kilometres-per-second – is hardly noticeable and the Doppler shift is difficult to detect.

Light passing through stardust does not appear noticeably more blue or red than normal, and the Doppler shift cannot be measured accurately. On the other hand, the stellar line spectra with sharply focused absorption lines have wavelengths that can be measured quite accurately, especially compared to known (at rest) values, and their Doppler shift can be measured easily and accurately. This shift in the line spectra can give the velocity of the emitting source and its direction of motion.

Over the next twenty years, Slipher would photograph spectra of more than 40 additional nebulae and would find the overwhelming majority to be receding (red-shifted) at speeds as large as 1,119 miles-per-second (1,800 kilometres-per-second). Slipher's observations of other spiral nebulae revealed radial velocities much greater than Andromeda, all of which were red-shifted, or moving away from the Earth. He did not realize at the time but he was observing, probably for the first time, the expansion of the universe. It is not surprising that he did not claim he had found evidence for an expanding universe in 1912. Much more evidence was necessary to draw this conclusion. To prove that the universe is expanding, it would be necessary to show that those objects located the farthest away from Earth have the largest receding velocities. Thus, distance and velocity would need to be measured simultaneously. This had to wait until the late 1920s when a larger telescope, the 100-inch (2.54-metre) reflector on Mount Wilson, became available.

The important thing is that by applying the simple Doppler effect, Slipher was announcing a new method for studying the stars and nebulae. Photographs taken of the line spectra of stars and nebulae indicated that the speed of these bodies were either receding or approaching the observer. In August 1914 Slipher reported at a meeting of the American Astronomical Society (AAS) that was held in Evanston, Illinois, that he had measured the spectra of thirteen galaxies, all of which were receding. At that time most galaxies were called spiral nebulae, and questions about their nature caused a raging controversy. Were they indeed external galaxies like our own Milky Way or were they nebulae, similar to the Orion Nebula that happened to have a spiral shape? No one knew. At that time astronomers did not even know that the Milky Way galaxy had a spiral shape.

What did Slipher's observations indicate? Slipher and the other astronomers of the time didn't know. However, they must have realized that these observations foretold something very important. According to astronomical legend, his 1914 presentation to the AAS produced one of the very few standing ovations in the history of the American Astronomical Society. One member of the audience, probably applauding with the rest, was Edwin Hubble, newly arrived at the University of

Chicago's Yerkes Observatory after deciding to become an astronomer.

Three years later, Slipher gave another paper that many consider his most important. By this time his observed ratio of red-shift to blue-shift had risen to 21:4, which meant that all but four of the twenty-five nebulae he measured were receding, moving away from the solar system. He added an astonishing interpretation, attributing a mean velocity of 700 kilometres-per-second for the solar system. He furthermore made a sensationally prescient point: For us to have such motion and the stars not show means that our entire stellar system moves and carries us with it.

Slither's life and career were remarkable for many reasons. In addition to the red-shift work, Slipher also discovered that these spiral nebulae are rotating. He also provided important data that helped to prove the existence of vast quantities of gas and dust in interstellar space. At the outset of his career, he had touched on planetary astronomy, Percival's passionate interest, and had started a spectrographic investigation of Jupiter, Saturn, Uranus and Neptune in 1903. Years after the founder died, Slipher supervised the search for Lowell's predicted Planet X.

Slipher spent his whole working life in Flagstaff, after being hired by Lowell himself in August of 1901. He became the assistant director of the Observatory in 1915, acting director upon Lowell's death in 1916 and Observatory director in 1926. He won many awards for his consistently careful observations over the years, and remained as director until his retirement in 1954. Fifteen years later he died in Flagstaff at the age of ninety-three.

The two seminal discoveries of 1912, namely Cepheid variable stars by Henrietta Leavitt and red-shift of nebulae by Vesto M. Slipher, were not made at Mount Wilson, but at Harvard and Lowell. G.E. Hale was still tooling up for MWO's coming four decades of pre-eminence in observational astronomy. Though the 60-inch (1.52-metre) telescope was in operation from 1908, the construction of the 100-inch (2.54-metre) telescope was only just beginning. But Leavitt and Slipher's discoveries would prove vital to the historic work by two astronomers about to arrive at Pasadena. Recruited by Hale to his observatory on the mountain in California, Harlow Shapley and Edwin Hubble were both born and raised in the state of Missouri and both started their professional training in fields other than astronomy.

Discovering The Milky Way Galaxy

One of the most striking features of life in the rural countryside in the twenty-first century is that the Milky Way is still visible in the night sky, an experience now virtually impossible for city dwellers due to the glare of ubiquitous artificial lighting. Appearing as a band of faint white line that stretches from one horizon to the other, the Milky Way stretches across the sky as we look at the billions of stars that make up the spiral galaxy that humans inhabit. Most inexperienced sky watchers are not impressed with what they see because it is so difficult to imagine the geometry from within this structure. Usually photos of other spiral galaxies similar to ours, like Andromeda, clarify the concept.

Consider how difficult this visualization was to early astronomers like William Herschel, who in 1785 first tried to picture our galaxy. In the northern hemisphere the band is brightest in the region of the constellation Cygnus and is best viewed in the summer. But in the southern hemisphere, the Milky Way is even brighter. Not bright enough, however, to be seen from urban areas. Indeed, many city dwellers have never even noticed the Milky Way in the sky.

In 1610, Galileo made the first telescopic observations on the Milky Way and discovered that it is composed of a multitude of individual stars. We now know that the Sun is located within this disc-shaped system of stars and that the Milky Way is the light from nearby stars that lie more or less in the plane of the disc. This great stellar system is called the Milky Way galaxy or, more simply, 'The Galaxy' – an anthrocentric term indicating that ours is the only galaxy in the universe, which was the accepted belief for centuries. This term can often cause confusion as astronomers have developed related words such as 'extragalactic', which means anything outside our own Milky Way galaxy.

In 1785 William Herschel showed that the system (that is to say, the Milky Way) to which the Sun belonged has the shape of a wheel or disc. He used the telescope to count the number of stars he could see in various directions. He found that the numbers were about the same in any direction around the Milky Way, a result that seemed to show that the Sun was near the centre of the galaxy. We now know that Herschel was right about the shape of the galaxy but wrong about the location of the Sun. Because of interstellar dust, which absorbs light from the stars and restricts optical observations in the plane of the Milky Way, Herschel was able to observe only a tiny fraction of the system of stars that surrounds us. Thus, until early in the twentieth century, astronomers accepted Herschel's conclusion that The Galaxy (the Milky Way) was centred on the Sun and extended only a few thousand light-years from it.

A major modification of this view was about to take place when in 1914, Harlow Shapley, a young man from Princeton University and the favourite graduate student of Henry Norris Russell received one of the most coveted positions a post-doc could imagine: An appointment to join the staff of the Mount Wilson Observatory and work under the tutelage of George Ellery Hale.

Harlow Shapley was born in November 1885, in the rural mid-west of the US, Nashville, Missouri. His childhood was spent on the family farm and he received much of his early education in a one-room schoolhouse. At the age of sixteen, Harlow took a course at a business school and became a newspaper reporter and city editor for the *Daily Sun* in Chanute, Kansas. He resolved to save his money and get an education. After attending the Carthage Collegiate Institute in Missouri, Shapley graduated as valedictorian of his class. He then entered the University of Missouri at Columbia, intent on studying journalism. After discovering that the journalism department wasn't yet open, he took up astronomy instead. (Later, Shapley would say that he had opened an alphabetical catalogue of courses, found himself unable to pronounce 'archaeology', and so went on to the next subject, 'astronomy'.) He earned his bachelor's degree in 1910 and his master's degree one year later. His academic career led him subsequently to Princeton, which was then being rejuvenated under the new administration of President Woodrow Wilson, with young Henry Norris Russell at the head of its astronomy department.

Shapley's arrival at Princeton could not have been timelier. Shortly before, Russell had embarked on a new approach to the analysis of the light curves of eclipsing binary stars in a quest of the properties of these variables. The arrival of a research student of Shapley's calibre was fortuitous both for the subject as well as for Russell himself. By 1912, thousands of observations were awaiting analysis. Russell was destined to become one of the leading lights of American astronomy, yet without Shapley's energy in applying new methods of analysis, the project would not have got off the ground. The systematic study that Shapley was assigned to would be the starting point for Russell's career-long classification of all stars for his influential HR diagram (see previous discussion). On the other hand, Shapley's work under Russell, which later became his 1914 PhD thesis on the orbits of 90 eclipsing binaries, created in virtually one stroke a new branch of double-star astronomy. During the project, Shapley developed a deep interest in reliable methods for determining the distances to stars.

After receiving his doctorate, Shapley travelled to Europe where he met many European astronomers. He then returned to Missouri to marry his sweetheart, Martha Betz, in April 1914. They had an unusual honeymoon, on top of a mountain in California. Fortune had smiled on Shapley (and on astronomy) when George Ellery Hale offered him a research post at the prestigious Mount Wilson, one of the few institutions in the US where such positions were available at that time. Shapley was thrilled to have access to the 60-inch (1.52-metre) telescope at MWO, and shifted his scientific interests from eclipsing variable stars to globular clusters, which are compact spheres composed of many thousands of stars. Studying these stellar configurations intensely for the next four years, he made important advances in the fields of Cepheid variables (he established for certain that they are pulsating stars rather than eclipsing binaries) and after that, the distribution in the sky of globular clusters. These studies of globular clusters would enable Shapley to complete what was to become his greatest single contribution to science: the discovery of the dimensions of our galaxy, the Milky Way, and of the location of its centre.

Before Shapley embarked on this journey, he had to be certain that the standard candle was an accurate calibration. Recall, the one thing that was missing from Henrietta Leavitt's work was an *absolute* distance measurement to the Cepheid variables in her sample. She had a period-luminosity relationship, but no calibration point for the relationship. It needed to be calibrated against something with a known distance. Hertzsprung had been the first person to try to calibrate this period-luminosity scale for Cepheid variables. Using his calibration, the distance to the Small Magellanic Cloud was estimated to be 30,000 light years (LY) away, much further than most expected (present-day estimates are closer to 170,000 LY).

The next attempt to calibrate the Cepheid scale came from Shapley himself, in 1918. He included Cepheids from globular clusters in his calibration of the period-luminosity relationship. This became the standard work on the subject. (It was this calibration of the relationship that Hubble used seven years later when he observed Cepheid variables in the Andromeda Nebula.) That the distribution of such clusters in the sky is highly asymmetric (most of them are situated in one hemisphere) was known long before Shapley's time. It was also known that many of them contain large numbers of Cepheid variables whose absolute brightness could be estimated from the periods of their brightness variations. Armed with this evidence, Shapley embarked on a systematic study of the distribution of the globular clusters in space. He found the centre of the distribution to be located some 50,000 light years from the solar system, in the direction of the constellation of Sagittarius. This point Shapley boldly identified as the centre of our entire galaxy. As a result, our galaxy began to emerge for what it actually is: a stellar system at least ten times as large as all earlier estimates, with our Sun located eccentrically some 50,000 LY from the galaxy's centre. (Modern methods have fixed this distance at 26,000 LY, but Shapley's prediction was correct to within a factor of two, no small achievement at the time.)

Shapley's estimate of the distance was, to be sure, somewhat exaggerated; by 1918 he was not yet aware of the absorption of light which a concentration of gas and dust close to the galactic place exerts on distant objects at low galactic latitudes. The role of this interstellar fog did not transpire more fully until the 1930s, and it reduced the estimate of the Earth's actual distance from the galactic centre from 50,000 to 30,000 LY; but the order of magnitude of Shapley's earlier estimate remained unaltered.

Naturally enough, Shapley's new, sensational results stirred up interest in the larger scheme of things and many re-evaluated their ideas on the structure of the entire universe. From Lord Rosse to Herschel and beyond, there had been disagreement on the nature of the so-called nebulae, the faint patches of light revealed in large numbers during the age of moderately large telescopes. Near the beginning of the nineteenth century, even the great Immanuel Kant professed the radical opinion held by many that the analogy of the nebulae to the system of stars in which we find ourselves (i.e. the Milky Way) is in perfect agreement with the concept that these are just 'island universes' – in other words, other Milky Ways.

Astronomical opinion rejected this idea for two centuries, and Shapley was no different. However, what shocked contemporary astronomers and provided a great deal of controversy was Shapley's measurement of the *size* of the Milky Way system. It was so large, Shapley said, that the system and all its globular clusters, which he used to find its centre, must indeed be all there was to the entire universe. He thought nebulae such as Andromeda were not independent systems as Kant suggested. At most, they were minor satellites of the Milky Way. It is interesting to note that throughout history, humans have had a difficult time accepting the immense size of the universe. This is true even up to the present day.

Not everyone supported Shapley's description and conclusions about our galaxy. Not too far from Mount Wilson, where Shapley had made his measurements, a keen observer at the Lick Observatory, Heber Curtis, would reject the idea that Cepheids could be used as reliable distance indicators. Curtis was convinced that the spiral nebulae were all galactic systems of their own located far outside the Milky Way and that our galaxy was much smaller than Shapley's estimates. In a complete rejection of these new ideas from Mount Wilson, Curtis suggested that the Milky Way was just a tiny fraction of the entire universe, and the solar system was at its centre – about as far away from Shapley's conclusions as one could get.

Curtis was sure that Andromeda was hundreds of thousands of light-years away, far outside our Galaxy in direct disagreement with Shapley's new results. Though he did agree with Shapley that the globular clusters fell outside the disc of the Galaxy, he did not agree they were off-centre nor as close to the solar system as Shapley maintained. The Lick people held the position that the clusters were located near the centre of the Galaxy, but much closer. Clearly, Curtis did not accept the Cepheids, which Shapley had used to position the globular clusters, as reliable distance indicators.

However, Shapley did also have some important support. If Andromeda was in fact a distinct galaxy that could be observed to be rotating as reports suggested, there would clearly be a violation of the universal speed limit, the speed of light. He used the example of a nova that had been observed in the Andromeda galaxy and had temporarily outshone the nucleus of the galaxy itself – a seemingly absurd amount of energy for a normal nova – to support his claims. He concluded that the nova and the galaxy itself must be within our own galaxy since, if Andromeda was a galaxy in its own right, the nova would have had to have been unthinkably bright in order to be seen from so far away.

On the other side of the argument, Curtis contended that Andromeda and other such nebulae were separate galaxies, or 'island universes', therefore siding with Kant. He showed that there were more novae in Andromeda than in the Milky Way. From this he could ask why there were more novae in one small section of the galaxy than the others. This led to his support for the idea that Andromeda was a separate galaxy with its own unique age and rate of novae occurrences.

The Great Debate

As these two proud men headed for a serious confrontation, C.G. Abbot, the Secretary of the National Academy of Sciences, introduced an idea that would provoke debate at the Academy's next meeting. He recalled that G.E. Hale had suggested at a council meeting late in 1919 that at the next Academy meeting (which was planned for the following April) an evening should be devoted to one of the annual lectures paid for by the fund set up in memory of Hale's father, William Ellery Hale. On 3 January 1920 the Home Secretary of the Academy wrote to Hale:

You mentioned the possibility of a sort of debate, either on the subject of the island universe or of relativity. From the way the English are rushing relativity in Nature and elsewhere it looks as if the subject would be done to death long before the next meeting of the Academy. Perhaps your first proposal, to try to get Campbell [Director of the Lick Observatory] and Shapley to discuss 'island universes' would be more interesting. I have a sort of fear, however, that the people care so little about island universes . . .

In spite of these reservations, the idea caught on, though Lick put forward Heber Curtis instead of Campbell. This pairing stoked the fires of controversy. Advertised as 'The Great Debate', in the end it was anything but a 'debate'. The arrangement became too formalized for a debate, and the event saw the two protagonists restating their positions in front of a distinguished crowd of astronomers. The focus was the nature of spiral nebulae and the size of the universe. The basic issues under discussion – representing the opposing views of Shapley and Curtis – were whether distant nebulae were relatively small and lay within our own galaxy or whether they were large, independent galaxies in their own right.

The 'debate' took place on 26 April 1920 in the Baird auditorium of the Smithsonian Museum of Natural History. The two scientists first presented independent technical papers about 'The Scale of the Universe' and then took part in a joint discussion that evening. Even Einstein was in attendance, in the middle of a whirlwind tour in America celebrating the results of the Eddington eclipse expedition. In the end much of the lore of the Great Debate grew out of two papers published by Shapley and Curtis in the May 1921 issue of the *Bulletin of the National Research Council*. The published papers each included counter-arguments to the position advocated by the other scientist at the 1920 meeting.

This event has become a celebrated event in the history of cosmology even though there was no clear winner of the 'debate'. Perhaps this is why it has become so important, focusing the differences of opinion on galaxies at a pivotal time, yet still leaving the matter open as science often does. It would later turn out that Shapley was correct about the Milky Way, but Curtis was correct about the nebulae.

To give some idea of the fervour with which Shapley tried to support his case for the nebulae, consider the famous story told by Hubble's research assistant Alan Sandage just after the great debate:

> Picture yourself during the early 1920s inside the dome of the 60-inch (1.52-metre) telescope on Mount Wilson. Milton Humason (who was trained by Shapley and later became Hubble's principal assistant) is talking to Shapley. Humason is showing Shapley photographs of stars he had found in the Andromeda Nebula – stars that appeared and then disappeared (suggesting that these are probably Cepheids outside the Milky Way). Shapley very patiently explains to his protégé that these objects could not be stars because the nebula was a nearby gaseous cloud within the Milky Way system. Shapley takes his handkerchief from his pocket and wipes the identifying marks off of the back of the photographic plate.

Hubble would show in 1924 that it was these very Cepheid variable stars in the Andromeda Nebula that proved it was a separate galaxy system.

Shapley was relentless in his pursuit of the Cepheid variables and by 1918 had plotted the period-brightness relations of 230 pulsating stars whose cycles ranged from 5 hours to 200 days. Then, after four years on the mountain, he published his results in several papers, making his claims about the size and makeup of the Milky Way galaxy. By now the man who would resolve these and a few other critical issues in cosmology – G.E. Hale's second vital recruit to his mountain observatory – had finally arrived at Mount Wilson to start his work. His name was Edwin Hubble and one could not find a more different character than Shapley.

The two men were cut from different cloth. Shapley, unsophisticated but brilliant in his pursuit of knowledge, still carried his Missouri accent while Hubble had cultivated a false, affected British accent which Shapley despised. Hubble, on the other hand, proud of his military experience and tradition, did not appreciate Shapley's pacifist tendencies. One more important difference would emerge between these two world-famous astronomers – it would begin to take shape after Shapley left Mount Wilson when Hubble got his hands on the Hooker telescope.

In retrospect the years 1914–1921, those that Shapley spent at Mount Wilson, mark the pinnacle of his scientific life. But Shapley did not distinguish himself only in research. He became well known as an outstanding lecturer and successful writer to a much broader audience – not only astronomers. In addition, Hale's example helped to develop Shapley's innate talents for administration. Small wonder that in 1920, after the untimely death of William Pickering, Shapley was offered the directorship of the Harvard College Observatory. He accepted and the observatory remained his academic home for the rest of his life.

Uncovering the Andromeda Galaxy

All these activities inevitably raised considerable demands on Shapley's time, but as busy as he was, he never abandoned active work in galactic and extragalactic research. After he left Mount Wilson, Shapley no longer had reflectors as large as those that he had used on the west coast for his research. His telescopes at Harvard possessed smaller apertures, but wider angular fields. It was with these that Shapley and his collaborators continued to explore the Magellanic clouds and to probe the spatial distribution of external galaxies.

The Second World War drew a close to this line of research. Once research started again in 1945, and astronomers were confronted by a new set of questions. In 1945, Shapley was sixty years of age and as an elder statesman of science, he was called upon to perform many extracurricular duties, including presiding over the American Astronomical Society, the American Association for the Advancement of Science and the National Society of Sigma Xi. All of these duties took much of his time from the observatory and from more creative work. As a result, the last seven years of Shapley's term of office as director of the Harvard Observatory were not, perhaps, marked by the same output as the decade preceding them and his observatory gradually yielded its leading position to others.

Shapley's retirement from the directorship in 1952 (he continued to serve as research professor until 1956) brought to a close an era which had kept Harvard Observatory in the forefront of astronomy progress for seventy-five years under the administration of two outstanding directors – a record unequalled in the annals of astronomy. Unlike his predecessor, E.C. Pickering, who died in office, Shapley survived his retirement for another twenty years. He continued to be an active lecturer and author for some time after 1956, but his creative output slowed down considerably. He died peacefully on 20 October 1972, just shy of his eighty-seventh birthday.

12

EDWIN HUBBLE AND THE EXPANDING UNIVERSE

The stage at Mount Wilson was now set for the arrival of the most enigmatic key figure in twentieth-century astronomy, Edwin P. Hubble. His life trajectory, from early schooling in Missouri to veneration in California via the city of Chicago and his beloved Great Britain, is worth following for its sheer bravado and the outrageous manner in which he stage-managed his own career.

As his story unfolds, it quickly becomes evident that Hubble changed our view of the universe more than any astronomer since Galileo. Even if we correct common misunderstandings about his achievements – that he was the person who first observed the high radial velocities of the galaxies or that he discovered the expansion of the universe – his list of accomplishments is astounding. In fact, though others initially made these two important discoveries, Hubble's drive, his scientific ability and communication skills enabled him to exploit these to the fullest. He was able to embrace the problem of the whole universe, make it his own and contribute more to our understanding of it than anyone before or since. Of course he did have the use of the largest telescope in the world and it became available to him at just the right time.

Hubble became the recognized world expert in observational astronomy. No one who read one of his books or heard one of his lectures, which described in simple yet vibrant terms his research and views on the universe, could ignore him. He had an exciting, compelling personality that was far different from the personalities of most astronomers and much more like those of the movie stars and writers who became his friends and companions in the later years of his life.

Hubble's early life has been the subject of some fiction. His official biography was written by a man who was his protégé, admirer and friend – meaning that the biography does not probe deeply into Hubble's early life. Other details about his life were accepted at face value, along with any embellishment provided by Hubble's widow, Grace. Grace in turn devoted years of her life to glorify the astronomer in a manuscript biography that is demonstrably false in many of its details, and that omits whole areas of his life and even more of her own.

More recent scholarship, particularly by Gale E. Christianson, one of the most respected contemporary historians of science, has given a 'warts and all' account of Hubble's life. In fact, in a review of Christianson's biography, *Edwin Hubble, Mariner of the Nebulae,* his character is described as 'one big wart'. This same reviewer also speculated that the sports promoter who wanted to train the young undergraduate Hubble to fight Jack Johnson – then the heavyweight boxing champion of the world – might have succeeded in taming the more unattractive aspects of Hubble's personality: 'For surely Johnson, the first African-American champion, would have pounded the vain, supercilious, obnoxious, pretentious, petty, egocentric, shallow, phoney, insensitive and very white Hubble into a quivering mass of protoplasm'.

Aside from critiques of his personality, it is important to understand the energy and skill with which Hubble approached science. By the end of his life, he had changed man's view of the universe as much as Copernicus and Galileo.

There is no question that Hubble's life was an interesting one. Born in November 1889 in Missouri, his father and mother could both trace their history through many generations of Yeoman families, and thus firmly belonged to a culture representing self-sufficiency and self-reliance, individualism and independence. His ancestors came from England, Ireland and Wales. The first family member to migrate to America, an officer of the royalist army, did so in the middle of the seventeenth century. During the Revolutionary war and the civil war his family members were soldiers; during times of peace, pioneers living on the land. They handed down their traditions, strong physique and qualities such as integrity, self-reliance, loyalty as citizens and loyalty to their families.

Edwin was the third of a large family of seven children. He started school in the town of Marshfield in 1895 and three years later his father transferred to the Chicago agency of his fire insurance firm and moved his family to nearby Evanston. Next, the family moved to a village west of Chicago, which was within commuting distance by train to his father's office in Chicago's Loop. As the grade school reports show, in eighth grade Edwin's average in different classes were typically a modest 85-90 per cent, but in high school he blossomed, particularly in English, mathematics and science, choosing to focus on chemistry, biology and physics. Throughout high school, Hubble was always one of the youngest students in his class, usually about two years younger than the average. He was only sixteen years old when he graduated at the end of the winter term in 1906.

In spite of his interest in science and mathematics, Edwin was not overly bookish. He led a healthy outdoor life with frequent summer vacations to his grandfather's dairy farm near Marshfield, and became a star high school athlete, especially on the track team. He was a high jumper, long jumper, shot putter and discus thrower and also ran on the relay team. His involvement in extra-curricular activities coupled with his excellent grades paid off for Hubble when, upon graduating, he was awarded a scholarship to the University of Chicago, which had opened only fourteen years earlier.

Edwin entered university as a sixteen-year-old freshman in September 1906 and again demonstrated his ability as a good student, particularly during the first two years. In addition to mathematics, chemistry, physics and astronomy he also selected French, Greek and Latin. At the end of his second (sophomore) year he received a two-year Associate in Science degree, as was then standard at the University of Chicago, and won a scholarship as the best physics student on his course. He proceeded to enrol in all of the advanced astronomy courses, mechanics and celestial mechanics – courses that would become important to him in his future career. He received his BSc degree at the end of March 1910 with enough credits to skip the spring quarter of his senior year. As he had done in high school, in spite of the fact that Hubble was two years younger than most of his classmates, he was a very good athlete and won varsity letters in track and basketball. In his junior year at Chicago, Hubble took the examination for the Rhodes scholarship, which would lead to the next important stage in his life.

In establishing these scholarships, British-born South African Cecil Rhodes wanted to strengthen the connection between the United States and the United Kingdom by bringing American students to study at Oxford University. Rhodes specified that each successful candidate should be a 'manly chap' who was a combined good student, athlete and leader. Rhodes scholars had to be between nineteen and twenty-five years old when selected and had usually completed their undergraduate degrees in America before they enrolled at Oxford. One scholarship was awarded in each state each year. Hubble was supremely qualified in every way. As if his record wasn't enough, Robert A. Millikan – a Nobel Laureate and later to be President at Chicago University – for whom Hubble had worked as a lab assistant in the elementary physics course, wrote a letter of recommendation for him. Millikan described Hubble as a man of magnificent physique, admirable scholarship with worthy and lovable character: 'I have seldom known a man who seemed to be better qualified to meet the conditions imposed by the founder of the Rhodes scholarship than Mr Hubble.'

Six Illinois candidates took the examination in December 1909 and, not surprisingly, Hubble was awarded the scholarship. At Oxford, he entered Queens College and read jurisprudence. Both his father and grandfather wanted him to become a lawyer and he was probably bowing to their pressure when he initially decided what to study. While at Oxford, he also became a passionate anglophile – adopting all manner of English affectation. A fellow Rhodes scholar later remarked that Hubble in one year had become very British, aping the language and manners of the Oxford upper crust. He managed to complete the jurisprudence course in the standard two years and received a second-class honours degree in 1912. He proceeded with his studies for a third-year, beginning the work required for a bachelor's degree in law. But no sooner had he begun his third year than he dropped law and switched to Spanish instead.

In 1913 he returned to the US, was admitted to the bar in Kentucky, and practised law for about a year in Louisville where he had no trouble passing the bar examination. Though most biographers include the fact that Hubble practised law for about a year in Louisville, there is absolutely no record in Kentucky that he ever did. What Hubble actually did was get a job teaching Spanish and physics in the high school at New Albany, Indiana, a suburb of Louisville just across the Ohio River. In addition to teaching, he coached the high school basketball team, leading his team through an undefeated season all the way to the state tournament in Bloomington. He was a very popular teacher and coach and the students in the class of 1914 dedicated their yearbook to their rugged, seemingly erudite Spanish and physics teacher.

It was at this point that Hubble realized that teaching was not for him, and that his passion lay instead with astronomy. He wrote to his astronomy professor at the University of

Chicago to ask about returning as a graduate student. He was immediately recommended to Edwin B. Frost, the director of Chicago's Yerkes Observatory, with the following note:

> Personally Hubble is a man of the finest type. Physically he is a splendid specimen. In his work here, altogether, and especially in science, he showed exceptional ability. I feel sure you would find him just the sort of man you would wish to have.

This recommendation sounded good to Frost, especially since Hubble had some experience with scientific instruments, notably, in the physics lab at the university, working with Millikan and others. At that time Yerkes needed graduate students who were well qualified to work as assistants on the observing programme with the 40-inch (1.02-metre) refractor telescope. Hubble was offered a service scholarship covering tuition and living expenses and wrote that he would come to Yerkes in the fall. Instead he arrived early, following Frost's advice that he attend the imminent meeting of the American Astronomical Society in August at nearby Evanston, Illinois.

This was the same meeting in which Vesto M. Slipher presented his important spectrographic observations of nebulae. Slipher had been the first to obtain well-exposed and calibrated spectrographs of spiral nebulae that not only showed their absorption line spectra but also revealed their significant Doppler shifts. These shifts implied large radial velocity, or motion away from the earth for most of the nebulae. Hubble was therefore part of the audience that gave a standing ovation to Slipher and must have felt the excitement in the air. Things were moving quickly for the would-be astronomer.

It was clear from Slipher's report that the Lowell Observatory was going to play a significant role in the future of American astronomy. By contrast, at Yerkes Hubble found a dying institution. Founded by G.E. Hale and built around the largest refracting telescope in the world at the time, Yerkes had gone into operation with great fanfare in 1897. However, eight years later, Hale had left permanently for southern California and taken with him outstanding astronomers such as George W. Ritchie, a master of instrument construction; Walter S. Adams, who would succeed Hale as director at Mount Wilson; and Francis Pease, who would become an expert observational astronomer. Frost, who inherited the Yerkes directorship, had at one time been a competent spectroscopist, but without any real creative ideas of his own. Frost continued the same routine programme, leaving it unchanged for years.

Hubble's choice could not have been timelier. The 24-inch (.6-metre) reflector built by Richie had been standing mostly idle since its maker left for Mount Wilson. Perhaps inspired by Slipher's paper, Hubble began a programme of photography with that telescope, noticing that definite changes had occurred within the nebulae in a few years since his previous work. The Yerkes programme grew into a PhD thesis entitled *Photographic Investigations of Faint Nebulae*, which remarkably contains many aspects of the later research on galaxies in the universe that he conducted at Mount Wilson. Though the thesis was not very good from a technical point of view, it did show the careful work of a promising scientist who knew how to manifest solutions to great problems.

This fledging work showed that Hubble already knew how to get the most out of a telescope once he understood the instrument's limitations. He knew the right questions to ask and had confidence in what he was observing and to use it effectively to explain his results. He stayed for a second year at Yerkes, took graduate courses in mathematics, physics and astronomy as well as celestial mechanics and the theory of applied differential equations. By this time he had clearly identified a strong inner desire to become a top astronomer.

In 1917 after completing his doctorate, he enlisted in the infantry and served in France as a line officer in the American Expeditionary Force. Apparently, Hubble was disappointed when he arrived in France during the First World War, only days before the November armistice.

As a student at Chicago and as a PhD candidate in astronomy at Chicago's Yerkes Observatory just before the war, Hubble had attracted the attention of the Director of the Mount Wilson Observatory, G.E. Hale. After the war, Hale followed up on an offer he made to Hubble for a staff position at Mount Wilson. But Hubble was so confident of himself and was in no hurry to leave England after the armistice, that he decided to stay and bask in the glory of his stature as a major in the US Army, part of the 'victorious' allied forces.

Recall that in the summer of 1919, Arthur Eddington was setting up his camera and telescope on an African island to photograph the eclipsed Sun to confirm Einstein's revolutionary gravitation theory. Meanwhile, back in Pasadena, Hale wrote to Hubble in England, pleading with him to return to America to take up a golden opportunity: 'Please come as soon as possible', the note read, 'we expect to get the 100-inch (2.54-metre) telescope into commission very soon, and there should be abundant opportunity for work by the time you arrive.'

Taking his own good time, Hubble sailed to New York some four months after Hale's request. From the east coast, he immediately took a train to San Francisco for his formal discharge from the military and then travelled the short journey to Pasadena. Still in uniform when he arrived at Mount Wilson, he was treated with extreme deference, no doubt to the chagrin of the Mount Wilson staff, and especially Harlow Shapley, who took an immediate dislike to the poseur with the English accent. Hubble, after all, originated from Shapley's native state of Missouri. His special observatory wardrobe of knickers, jodhpurs, high-topped military boots and Norfolk jacket irritated Shapley to distraction.

Less than two years later, Shapley became the director of the Harvard College Observatory. After plotting the period-luminosity curves for 230 pulsating Cepheids, he was pleased that his model of the universe, which consisted of only the Milky Way with an off-centred Sun, would last forever. But once Hubble got the use of the powerful 100-inch (2.54-metre) telescope,

Shapley's model of the universe was bound to change. It would last less than five years before a new era began.

Hubble arrived at Mount Wilson at the ideal time. The 60-inch (1.52-metre) reflector telescope had been in use since 1908 and the 100-inch (2.54-metre) Hooker telescope was first pointed at the sky in 1917, but was then stalled by the war. On 11 September 1919, just a few days after Hubble's arrival on the mountain, the Hooker telescope was finally opened to the staff. Fortunately, from his Yerkes experience, Hubble already knew how to use a reflector effectively. He also knew how to photograph nebulae; and so lost no time getting down to work. Milton L. Humason, who would later become an outstanding staff member and Hubble's collaborator in studying the expanding universe, was given the task of initiating the newcomer. On 25 October 1919, the first time he used the Mount Wilson telescopes, Hubble made a distinct impression on Humason:

> I received a vivid impression of the man that night that has remained with me over the years. He was photographing at the Newtonian focus of the 60-inch (1.52-metre) telescope, standing while he did his guiding. His tall, vigorous figure, pipe in mouth, was clearly outlined against the sky. A brisk wind whipped his military trench coat around his body and occasionally blew sparks from his pipe into the darkness of the dome. We estimated visibility that night as quite the poor on our Mount Wilson scale, but when Hubble came back from developing his plate he was jubilant. 'If this is a sample of poor seeing conditions', he said, 'I should always be able to get usable photographs with the Mount Wilson instruments'. The confidence and enthusiasm which he showed on that night were typical of the way he approached all his problems. He was sure of himself – of what he wanted to do and how to do it.'

For all of his pretensions, Hubble was certainly not a fraud. Once he got to the controls of the Hooker telescope, he knew just what to do to try to make sense of the cloudy patches in the sky (the nebulae). The French astronomer Charles Messier had catalogued these celestial objects in the eighteenth century and thereafter each nebula was denoted by a letter M (for Messier), followed by a number. For example, M31 is the great galaxy in Andromeda and M51 is the Whirlpool Galaxy.

If these objects were nearby, then the cloudy patches were luminous clouds of gas within the Milky Way. On the other hand, if they are very remote and far beyond the stars of the Milky Way, they could be massive systems of billions of stars which simply have the appearance of cloudy patches because they are otherwise too remote for individual stars to be resolved. In this case, they would be considered galaxies in their own right, so-called 'island universes'. Hubble's task was to use the tremendous power of the Hooker telescope to locate and resolve a few individual Cepheid variables within these nebulae. He could then measure their periods of luminosity variation. He would then calculate their distance using the technique that had been initially investigated by Henrietta Leavitt and further refined by Shapley.

With superb technique and great patience, Hubble found variable stars in M31, M51 and another 'nebula' classified as NGC 6822. During 1923–24 he analysed the light curves of these variables and found that they were indeed Cepheids. Since this type of star was in a class of super giant stars, and the ones studied by Hubble were very faint, this gave him a clue that they were probably in systems far beyond the boundaries of the Milky Way.

On the night of 4 October 1923 Hubble focused in on a spiral arm of M31 (Andromeda). Visibility was not good. Yet despite the poor conditions, a 40-minute exposure yielded a photograph showing a suspected nova, which Hubble intended to confirm the next night. (A nova is an exploding star that changes brightness dramatically over a period of just a few days and might be confused with a Cepheid.)

The next night, under improved conditions, Hubble confirmed his suspicion that the star was indeed a nova. But a closer examination of the exposed photographic plate revealed two more stars that Hubble concluded were also novae. After completing his run, he anxiously travelled down to the observatory offices on Santa Barbara Street in Pasadena. Once there, he searched the files for previous photographs of the same part of the sky. He was shocked to find that the star on the plate was not a nova at all, but a Cepheid variable. (This plate, H335H, was destined to become the most famous photograph ever taken on Mount Wilson).

Hubble was then able to estimate the distance to M31, the so-called Andromeda Galaxy, as almost 1,000,000 LY from the earth. Since Shapley had estimated the diameter of the entire Milky Way to be approximately 300,000 LY, it was quite clear that Andromeda was an island universe, well away from, and certainly not part of, the Milky Way. As legend would have it, Hubble took out his marking pen and near the top of the famous photographic plate, crossed out the 'N' that was used to denote 'nova' and in bold capital letters printed 'VAR' for 'variable', followed by an exclamation point. The known universe had expanded dramatically that day and – in a sense – the cosmos itself had been discovered! The *New York Times* realized the importance of the discovery and that day printed a headline which read 'Astronomer Finds Spiral Nebulae Are Stellar Systems'.

Following the Andromeda galaxy work, Hubble tried to define a scheme for understanding the formation and evolution of galaxies, and proposed what is now known as the tuning fork diagram. This diagram was a galaxy classification scheme in which the handle of the tuning fork consists of an evolving sequence of elliptical galaxies (from spherical to true elliptical), with one arm of the tuning fork diagramming the evolution of spiral galaxies, and the other arm of the tuning fork diagramming the evolution of the 'barred' spiral galaxies. This scheme did not impress the astronomical community. However, his next project was different. Hubble used the Hooker telescope to follow-up the work started by Vesto Slipher, measuring the red-shifts of distant nebulae. Milton Humason, a veteran of the Mount Wilson staff who would later become a legend up on the mountain, assisted him in this systematic study.

Milton Lasell Humason rose from humble beginnings to become a top American astronomer. Born in Dodge Center, Minnesota of middle-class parents who later moved to California, he dropped out of school and had no formal education past the age of fourteen. His love of mountains brought him to Mount Wilson, where he took a job as a mule driver for the pack trains that carried construction equipment up the mountain from the Sierra Madre. This happened during the building of the large observatory telescopes under the directorship of George Ellery Hale. In 1917, after a short stint on a ranch in La Verne, he became a janitor at Mount Wilson, often working late at night.

One evening when a staff member did not appear as scheduled, Humason volunteered to be a night assistant at the observatory. He continued to offer his services whenever needed and soon his technical skill and quiet manner made him a favoured assistant. Hale soon recognized his talent and the innovative director made him a Mount Wilson staff member in 1919.

This was unprecedented, as Humason did not have a PhD, or even a high school diploma, but he soon proved Hale's judgment correct and assisted Harlow Shapley during his work on the structure of the Milky Way. In exchange, Shapley taught Humason how to manipulate the large telescope and adjust the controls to get the best results for each measurement. Humason became known as a meticulous observer and after Shapley left, began to work with Edwin Hubble. This was an ideal partnership, as Humason became skilled at obtaining photographs and difficult spectrograms of faint galaxies.

Humason took part in an extensive study of the properties of galaxies that was initiated by Hubble in 1928. The work consisted of making a series of systematic spectroscopic observations to test and extend the relationship that Hubble had found between the red shifts (which yield the velocity) and the apparent magnitudes (which yield the distance) of the galaxies. But because of the low surface brightness of galaxies there were always severe technical difficulties. Humason personally developed a technique for determining the exposures and plate measurements for these difficult images and soon became the master of the giant Hooker telescope. By the late 1920s, Humason was collaborating with Hubble by photographing the spectra of faint galaxies with the telescope.

Earlier in the decade, Hubble had made some preliminary determinations of velocity versus distance relations, and was able to boldly draw a straight line through the scattered points. This linear relationship is indicated in the graph on the left below. He published these results in 1929 in the *Proceedings of the National Academy of Science* in a paper he titled 'A Relation Between Distance and Radial Velocity Among Extra-Galactic Nebulae'. In his conclusions, he confirmed that not only is the universe expanding *continuously*, but that it is expanding *uniformly*.

By the time of this discovery, Humason had become Hubble's full-time assistant. Not one to be unduly interested in the credentials (or the lack of) of his collaborators, Hubble's interest in Humason was grounded in the skills that he had developed in working with the large telescopes. Hubble and Humason would form a formidable partnership and spend many frigid nights on the mountain steering the giant Hooker telescope across the heavens, locking on to elusive objects of interest before finally photographing the faint images. Their images of the spectra and brightness of the stars and nebulae would change man's view of the cosmos overnight.

Their very first joint measurement was historic. During two freezing nights on the mountain, focusing on a faint nebula that Slipher was unable to resolve, they produced a plate with a very large red shift, confirming a speed of recession of 1,864 miles-per-second (3,000 kilometres-per-second), some 1,118 miles-per-second (1,800 kilometres-per-second) greater than Slipher's largest value. This was a great moment for Humason and his mentor. In addition, it produced a new and welcome data point on the velocity-distance graph, indicating that Hubble's initial guess of linearity was correct. The universe was expanding uniformly, and now they had proof.

There was more to come. By 1931, Hubble and Humason had jointly published their classic paper in the *Astrophysical Journal*. In this paper they compared distances and velocities of remote galaxies moving away from the Earth at speeds as high as 20,000 kilometres-per-second as shown in the graph on the right (this is about 7 per cent of the speed of light). By extending these measurements well beyond what Hubble alone had initially published, they had demonstrated that the velocities of recession of galaxies are proportional to their distance from the Earth. Written as an equation, the relationship is linear, a simple straight line graph of velocity-versus-distance:

$$V \text{ (velocity)} = H_o \text{ (constant)} \times D \text{ (distance)}$$

Where V is the recession speed; D is the distance and H_o is a pure number called the Hubble constant. This equation is now known as Hubble's Law. Another way of expressing this relation is to say that the speed with which galaxies recede from an observer is proportional to the distance the galaxy is from the observer.

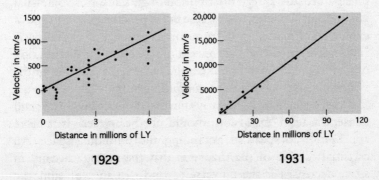

Measurements of Hubble's Law: 1929 & 1931.

The fact that galaxies obey this law shows beyond a doubt that the universe is expanding uniformly. Such a universe has two distinct characteristics: First, it is expanding at the same rate everywhere according to all observers in the universe, no matter where they are located; and secondly, astronomers will observe a proportionality between the velocities and distances of remote galaxies; that is, for a uniform expansion, Hubble's Law must be observed.

The concept of a universe expanding at the same rate everywhere is difficult to absorb. However, the concept can be made easier if we draw on a metaphor. Consider a loaf of raisin bread in which the cook may have put too much yeast in the dough. If the bread is set out to rise, it doubles in size during the next hour and the raisins move farther apart. If we select one raisin in particular, it is seen that since each distance doubles during the hour, each raisin must move away from the selected one at a speed proportional to its distance. The same is true no matter which raisin is chosen at the start.

But the analogy must not be carried too far; in the bread it is the expanding dough that carries the raisins apart, but in the universe there is no pervading medium presumed to separate the galaxies. Since 1905, Einstein has shown that there was no need to postulate a substance that pervades all of space. The existence of an expanding universe implies that the cosmos has evolved from a dense concentration of matter into the present broadly spread distribution of galaxies. It is somewhat misleading, however, to describe the expansion as some type of explosion of matter away from some particular point in space. In Einstein's universe the concept of space and the distribution of matter are intimately linked; the observed expansion of the system of galaxies *reveals the unfolding of space itself*.

Thus, even though the early universe was extremely hot and dense, the term 'explosion' should not be applied. It is more appropriate to speak of expanding and unfolding space. An essential feature of the theory is that the average density in space decreases as the universe expands. In an explosion, the fastest particles move out into empty space, but in Einstein's cosmology, particles uniformly fill all of the space. The expansion of the universe has had little influence on the size of galaxies or even the clusters of galaxies that are bound by the local gravity. Space is simply opening up between them.

In this sense, the expansion *is* similar to a rising loaf of raisin bread. The dough is analogous to space, and the raisins to clusters of galaxies. As the dough expands, the raisins move away from one another but stay intact. Moreover, the speed with which any two raisins move apart is directly and positively related to the amount of dough separating them. The analogy even holds up when considering the answer to questions about why the expanding universe does not pull the galaxies apart or increase their size, that is, because local gravity keeps the galaxies and clusters intact. Similarly, the raisins stay intact, yet move away from one another uniformly.

The three-dimensional raisin loaf does allow us to think of the medium of space (dough) expanding and the distances between the galaxies (raisins) increasing, sometimes at very high rates. It should be clear that if the universe is uniformly expanding, all observers everywhere including earthbound humans, must see other objects moving away from them at speeds that increase in proportion to their distance of separation. Hubble and Humason demonstrated that this is precisely what observers on Earth see when they graphed the Hubble Law.

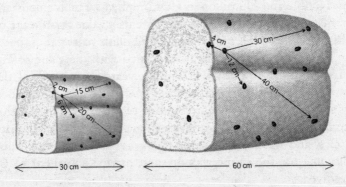

Raisin bread analogy of the expanding universe.

Though guided by the vision and insight of his mentor, Humason was soon making most of the exposures and plate measurements himself. Eventually, the velocities of 620 galaxies were measured and the results, published in 1956, still represent the majority of the known values of radial velocities for normal galaxies. The two astronomers worked together for many years and much of the work Humason performed was actually credited to Hubble. However, this was not Hubble's, decision, but the work of journalists and publicists. In spite of his penchant for self-promotion, Hubble gave Humason full credit as a co-author in the most famous of their papers about the expanding universe.

Hubble's law, a simple relationship describing the velocity-distance relation for the galaxies, had a very important implication for the age of the universe. V (velocity) = H_o (constant) x D (distance), the equation of Hubble's law, can yield the maximum age of the universe. An experimental plot of V (velocity) = H_o (constant) x D (distance) is the equation of Hubble's law. This result is a straight-line graph with slope = H_o, yielding the maximum age of the universe. The age of the universe is simply the reciprocal of the number H_o, the slope of the straight line plot of V versus D.

The first data with red-shifts up to 685 miles-per-second (1,100 kilometres-per-second) reported by Hubble in 1929, and by Hubble and Humason in 1931, implied a Hubble constant of 500 +/- 10 per cent. This value yields a value for the age of the universe = $1/H_o$ of just about 1.9 billion years.

There were two problems associated with the value of H_o. First of all, the value of 1.9 billion years for the age of the universe was smaller than the age of the Earth as determined at that time by the radioactive dating of minerals in the Earth's crust. This technique had just become respectable in the late 1920s, and was thought to be highly reliable. This was troublesome for the Hubble camp, and needed somehow to be resolved.

The second problem was the wide variability of the reported

values of H_o as determined by different investigators – sometimes from the same data. This variation was due to the arbitrary choice of the slope of the line (which determines H_o) through the experimental points. The measured values of the velocity and distance of far away galaxies are not very precise and estimates are often quite subjective.

Clearly, the Earth cannot be older than the universe, which means that one of these age estimates must be wrong. Since the radioactive dating of the Earth's crust had been reliably cross-checked on different elements, the finger was pointed at the hairy business of determining Hubble's constant, H_o from the velocity-distance graphs. This problem, which soon became known as the timescale difficulty, gained notoriety in the 1930s as the 'nightmare of the cosmologists' as described by historian of astronomy, John North.

As an accurate method for dating objects on the earth, radioactivity gained respectability in the early part of the twentieth century, and by the 1920s it had a reputation for yielding reliable estimates. In 1921, H.N. Russell, by now an authority on American astronomy, used thorium and uranium compounds to estimate the maximum age of the Earth's crust to be approximately 4 billion years. By the 1930s, Ernest Rutherford had found from the half-lives of uranium isotopes U-235 and U-238 that the upper limit of the age of the Earth was about 3.4 billion years. This was the best estimate of the average age of the Earth at that time based on radioactive techniques.

Cosmologists were clearly getting nervous about discrepancies thrown up by the radioactivity data, which was about a factor of two larger than Hubble's results for the age of the universe. Consequently, the timescale difficulty was often a priority for discussion at meetings about expanding-universe cosmology.

In 1931, whilst Einstein was a visiting professor at CalTech, he visited Hubble at Mount Wilson, examined the data and admitted that the introduction of the 'cosmological constant' was the biggest blunder of his life. Hubble's work convinced

Einstein that the universe was expanding and thus rendered his concept of the cosmological constant useless. A story often told about the 1931 visit reveals how the German scientist felt about Hubble's observations. Hubble's wife Grace often had the task of driving Einstein around the CalTech campus during his visit. One day, as the story goes, Einstein wistfully disclosed to Grace that he thought her 'husband's work is beautiful'.

Less than two years later, Einstein was back in Pasadena. This time he had more on his mind than the cosmological constant. In December 1932, he had left Germany on the Oakland steamer with thirty pieces of luggage, not sure that he would ever return to Germany as National Socialism, with its strong anti-Semitic programme, was gaining power. Travelling with him from Europe was the Belgian cosmologist Abbé Georges Lemaître. Together, they would give a series of seminars at CalTech in January 1933. Astronomers from Mount Wilson, including the notable Edwin Hubble, would attend these lectures.

Lemaître would present his theory of the expanding universe. Based on Einstein's theory of general relativity, Lemaître described the possibility of a universe that expanded from its initial state as a single point, a primeval atom about the size of the solar system. Lemaître contended that before the expansion began, the universe did not exist and moreover that the universe has a specific time at which it was created. The Belgian therefore made this prediction before Hubble's defining observations were announced. After Lemaître detailed his theory, Einstein famously stood up, applauded, and said, 'This is the most beautiful and satisfactory explanation of creation to which I have ever listened.'

Shortly after the excitement of Lemaître's seminars in Pasadena, Einstein recognized a problem that Lemaître's calculated rate of expansion had not resolved: If the universe was expanding at a steady rate, then the time it had taken to cover its radius was too short to allow for the formation of stars and planets. Lemaître solved this problem by expropriating

Einstein's cosmological constant. Where Einstein had used it in an attempt to keep the universe at a steady size, Lemaître used it to speed up the expansion of the universe over time. Einstein was clearly worried about the contradictions:

> The age of the universe must certainly exceed that of the firm crust of the earth as found from the radioactive minerals. Since the determination of age by these minerals is reliable in every respect, the cosmological theory here presented would be disproved if it were found to contradict any such results. In this case I see no reasonable solution.

It appeared that the father of relativistic cosmology and a strong believer in the supremacy of theory over observational facts was ready to give up his theory because of the timescale discrepancy.

It was not until 1936 that Hubble felt comfortable enough to express his concern over the contradictions that the timescale difficulty represented to relativistic cosmology. He concluded that perhaps the distance measurements were unreliable or the red shifts did not really represent the expansion of the universe. When a possible way out of the dilemma appeared, that is in questioning the value of the Hubble constant, confidence in the respected Mount Wilson astronomer and the 100-inch (2.54-metre) telescope was such that this line of questioning did not survive amongst astronomers. Hubble was considered a safe source and it was generally believed that the measurements had been carried out correctly. For the time being at least.

The difficulty was finally resolved by adjusting the experimentally determined value of the Hubble constant. Velocity-distance graphs depend on the measured values of velocity from the red-shifts of the spectral lines and the distances derived from the brightness of the Cepheid variable stars. The distances were wrong. In fact, a major adjustment to the distances used by Hubble on the M31 nebula in his studies reported in 1929 and 1931 was required and did not take place.

A recalibration of the Cepheid variable stars that Hubble (and Shapley before him) had used to determine distances in their pioneering studies was carried out. In a monumental body of work lasting over twenty years, it was discovered that the absolute brightness of the Cepheid variable stars in M31 was four times greater than the values used by Hubble. This correction resulted in a major increase in the inferred distance to M31 from 1 million LY to 2 million LY. Eventually, the theory was adjusted and shown to be consistent with the radioactivity data. But it did take some time.

Two Types of Stars

The man who carried out this work, another hero of modern cosmology, was drawn to the Mount Wilson Observatory from Germany. Walter Baade began his career on the famous mountain and eventually joined G.E. Hale, Harlow Shapley, Edwin Hubble and Milton Humason in the tradition of great astronomical observers of the twentieth century. Baade arrived in 1933, the same year Einstein and Lemaître had their epoch-defining visit with Hubble to discuss the expanding universe. Born and educated in a small town in northwest Germany, Baade's father was a schoolteacher. As a child Baade was an excellent student and learned to speak six languages including English, Latin and Hebrew as well as his native German. He always felt that his parents wanted him to be a clergyman, but his interest and ability in physics and mathematics directed him completely and unequivocally into science.

After attending gymnasium (the equivalent of a public school in Britain), Baade enrolled at Göttingen University, one of the leading European universities. A decade later, Göttingen would become the centre of research in quantum mechanics and attract theorists from all over the world. Despite learning his mathematics from two of the world's leading scholars, Felix Klein and David Hilbert, Baade was interested in observational astronomy. Immediately attracted to problems associated with astrophysics, he selected for his thesis the study of the spectrum of ß Lyrae, a bright eclipsing binary star system for which spectroscopic plates had been taken at Göttingen from 1900 to 1909. He proceeded to measure, reduce and analyse the plates with intensity and passion.

This was the beginning of his interest in developing the skills to extract information from spectrograms, which would shape his entire career. He accelerated his studies and received his PhD in astrophysics in 1919. This was the year Eddington proved Einstein's theory of general relativity with the eclipse photograph and was also the same year that the Mount Wilson 100-inch (2.54-metre) telescope went into operation.

Baade was aware of the exciting developments in California and immediately tried to go to America to work with the big telescopes. However, this was not possible for any German citizen after the First World War. Instead he took a job at the Hamburg Observatory and worked on the 40-inch (1.02-metre) reflector, the largest telescope in Germany. Continuing his interest in spectroscopic studies, he soon became an expert observer and it was not long before he was placed in charge of the telescope. A few years later, on 24 January 1925, fate played a role in Baade's life. A total solar eclipse was due to take place off the northeastern coast of the North American continent, and Baade was selected to observe the eclipse from a ship in the North Atlantic. After the eclipse, the ship put him ashore in the northeastern US, not far from Boston and Harvard University.

He was fortunate to have met Harlow Shapley in Germany in 1922 and had since corresponded with him on globular cluster research. Shapley liked Baade's friendly manner as well as his outstanding technical ability. From his powerful position as Director of the Harvard College Observatory, he helped the German obtain a Rockefeller Fellowship to spend a year in the US during 1926–27. Baade took advantage of this year in the US, visiting Harvard, Yerkes, Lick and Mount Wilson where he spent most of the year. At MWO during the time when Hubble was discovering the expansion of the universe, he did some outstanding work and learned much about how to use the big telescopes.

When Baade returned to Hamburg, the university officials immediately promoted him, ostensibly to keep him from attempting to migrate to America. He continued his research studying globular clusters and discovered several small faint clusters of galaxies. He knew that investigating these faint galaxies would be the future of studies in astrophysics and he yearned to use the great machines in America to this end. But by now, Baade had published his studies in important journals and had become internationally known. As a result, in 1931 he was offered a position on the Mount Wilson Observatory staff, which he immediately accepted.

Baade was very comfortable with the American approach to science practised at Mount Wilson, and was popular with his colleagues. He plunged into research on galaxies and globular clusters with the big telescopes. He chose to stay away from the 'cosmological problem', which was clearly in good hands with Hubble and Humason. With his time on the 100-inch (2.54-metre) Hooker, he carried out superb research on supernova and galactic structure. He spent much of his time trying to improve methods to accurately measure the brightness of very faint objects by photographic photometry, which was in its early stages at that time.

When the Second World War broke out, Germany and the US were once again at war. As a result, Baade was the one Mount Wilson astronomer who could not work on wartime technical developments since he was a German alien. He was also subject to wartime curfew and was not allowed on the mountaintop after dark. Only because MWO's Director, Walter Adams, pleaded to the highest levels of the American military bureaucracy was Baade exempted from this regulation and able to take advantage of the greatest opportunity of his career.

Not only did Baade receive maximum viewing time on the big telescopes as the younger astronomers were drafted into the service of America's war effort, but there was another advantage for him during the war years. 'Brown outs', in which external lighting all over Los Angeles was reduced, improved viewing from Mount Wilson immensely. These enabled Baade to obtain photographic plates of quite exceptional quality, identifying not only the brightest new stars in the centre of M31, but also the faint stars in the galaxy as well.

This was the opportunity Baade had been preparing for the duration of his life as an astronomer. With great care and skill he succeeded in resolving many individual stars in the nuclear bulge of the spiral galaxy M31 and noticed an amazing similarity between the stars in the nuclear bulge of Andromeda, those in the globular clusters and the stars in the halo of our own Milky Way galaxy. Realizing these stars were different from all the stars found in the spiral arms, he postulated that there are two different types of stars in the universe. He defined the red stars seen in the centre as population I and the bright blue stars in the spiral arms as population II. He reported the important general discovery that different spectral classes of stars, that is, stars of different brightness and colour, form the different groups or 'populations' within spiral galaxies. His two papers of 1944 describing this discovery made him famous among astronomers the world over.

At the time, Baade did not understand the physical cause of the difference between populations I and II. However, he did receive an unusual note from another émigré working on cosmology in the US, the Russian George Gamow, who wrote to him immediately after his papers were published. Written on a postcard sent to Baade at Mount Wilson, Gamow stated simply and emphatically that population I stars are *young* and population II stars are *old*. Baade gradually came to accept Gamow's idea, reiterating it as the physical meaning behind the two stellar populations at an important talk given at the 1952 International Astronomical Union General Assembly in Rome.

One important consequence of Baade's studies was his

discovery of a significant difference in the absolute luminosity of Cepheid variables for the two populations, even though they had the same period of pulsation. This held implications for the accepted cosmological distance scale since Harlow Shapley and Edwin Hubble had both assumed that the Cepheid variable stars in the globular clusters in our galaxy had the same characteristics as those in M31. Later, with the 200-inch (5.08-metre) telescope at Mount Palomar, Baade further discovered that with the same period of pulsation, population I stars are four times brighter than population II stars, indicating that the distance to M31, and other extragalactic distances needed to be increased twofold. So, M31 was not 1,000,000 LY away but really 2,000,000 LY away. Baade had doubled the size of the universe in a single stroke.

Reporting these results after the war at an important conference in Rome in 1952, Baade announced that everyone had overestimated the value of Hubble's constant because the accepted distance to M31 was (about) a factor of two too small. His adjustment reduced the constant, H_o, to one-half the result reported by Hubble. Consequently, the inferred age of the universe was increased by a factor of two to about 3.80 billion years, safely larger than the estimated age of the Earth from the radioactive dating of the earth's crust. Though this was not the final word on adjustments to Hubble's constant, cosmologists could breath a sigh of relief regarding the timescale difficulty.

The constant has been lowered constantly, beginning with Hubble in 1929 and continuing almost to the present day. An important historian of astronomy, Virginia Trimble, emphasized this in an entertaining paper in 1966 titled 'H_o, The Incredible Shrinking Constant, 1925–1975', in which she summarized the changes in H_o as decreasing by continuous adjustment of the data over a fifty-year period.

A few years after meeting in Rome, in 1956 Humason and Alan Sandage published red-shift measurements for 474 galaxies. With this data, Hubble's constant was reduced again, with the age of the universe now increased to 5.6 billion years. These revisions, which began with Baade's work, finally eliminated the discrepancy between the age of the universe and the age of the Earth from radioactive measurements. The timescale difficulty was gone.

By the end of the Second World War, Baade had become a key part of the American astronomical establishment and one of the leading figures of contemporary research. He was one of a very few dignitaries bestowed with the honour of presenting an address at the inaugural ceremony at Mount Palomar in June 1948, in which the new telescope was named after George Ellery Hale. On that occasion, Baade outlined how the programmes of the combined facilities at the Mount Wilson and Palomar Observatories would be applied. His talk was titled: 'A Program of Extragalactic Research for the 200-inch Hale Telescope.'

Baade was only four years younger than Hubble, who had been a member of the Mount Wilson staff for twelve years before the German arrived. From the beginning, Baade was therefore considered a junior astronomer to the world-famous discoverer of the expanding universe. The two astronomers were quite different in their approach to their work. After about 1931, Hubble was working almost entirely on the cosmological problem, which Baade avoided completely, choosing instead to work on globular clusters and galaxies as individual entities.

Their personalities were also entirely different. Hubble projected an imposing public image, courting personal publicity outside the circles of his work and did not have many friends among astronomers. Hubble's selfish and despicable *modus operandi* contrasted starkly with Baade's. For example, Hubble hired a publicist to get himself on the cover of *Time* magazine and when Albert Einstein visited Mount Wilson, he stayed close to Einstein to help ensure that he would be photographed with the great physicist. During his life, Hubble had tried to obtain the Nobel Prize, even hiring a publicity agent in the 1940s to help promote his cause, but all the effort was in vain as there was no Nobel category for astronomy. He was, however, awarded the Medal of Merit in 1946.

Baade, on the other hand, projected a personable, friendly, approachable image, especially with his colleagues on Mount Wilson. He disliked and distrusted personal focus in the outside world and tried to hide from it. But he never criticized Hubble and seemed content to take the role as the junior and less famous contributor to astronomy from his post at Mount Wilson.

The meeting with Einstein and Lemaître in 1933 more or less coincided with the beginning of the end of Hubble's era of pre-eminence in cosmology. The old guard of astronomy was disappearing. Though he worked on indefatigably at Mount Wilson until the summer of 1942, Hubble made no more great discoveries and left the mountain to serve in the Second World War as a technical adviser at the Aberdeen Proving Ground, a United States Army facility located near Aberdeen, Maryland.

As a farewell gesture, Hubble returned to the observatory after the war and made his last contribution to astronomy as a key consultant on the design and construction of the Hale 200-inch (5.08-metre) telescope on Palomar Mountain. In 1949, he was honoured by being the first to use the new telescope. Later that same year his first heart attack left him tired and sick. A second severe heart seizure ended his life in 1953, just as he was preparing for several nights of observations at Palomar.

The Hubble era was over. However, the heroic astronomer with the pretentious British accent had certainly left his mark. Today his name is associated with many things in astronomy and cosmology: the Hubble galaxy type, the Hubble sequence, the Hubble luminosity profile for nebulae, the Hubble constant (of course), the Hubble time, the Hubble radius for the universe and the most well-known use of his name, the Hubble Space Telescope. There is no doubt that his work should be celebrated in the same context as those great astronomers from history: Copernicus, Tycho, Kepler, Galileo and Newton.

Alan Sandage, Hubble's assistant, summed up Hubble's contribution to astronomy in an article published in the *Journal of the Royal Astronomical Society of Canada* in December 1989. Sandage claimed that from 1922 to 1936 Hubble solved four of the central problems in cosmology, any one of which would have guaranteed him a position of importance in the history of astronomy.

Hubble would have been thrilled to know that the Space Telescope is named after him. His legacy has ensured that astronomers continue to fulfil the 'hope to find something not expected', that he once articulated in a 1948 BBC broadcast from London.

13

THE BIG BANG

During the quarter of a century preceding the Second World War, from the moment Einstein first published his field equations, major developments occurred in the application of these magical equations to the structure of the universe. Theoretical difficulties consisted of finding a simultaneous solution to the set of equations that would also be appropriate in a universe that human beings actually inhabited. Quite apart from Einstein's miraculous achievement of discovering the field equations, it was necessary to impose certain acceptable conditions that might render the mathematics manageable and useful.

Einstein himself made the first attempt. He found that if he assumed that the universe contained any matter at all, the solution he obtained was unstable, suggesting that the universe would be in the state of uncontrolled expansion or collapse. This contradicted prevailing reports from astronomers who contended that the universe was static. It was at this stage in the conundrum that the infamous cosmological constant, or 'fudge-factor', was introduced to re-balance the equation and produce a static solution.

The Dutch astronomer Willem de Sitter was next to try to find a solution to the equations. He succeeded, but had to postulate an empty universe, which contained no mass at all. Otherwise, he found that the gravitational attraction of any mass would cause the universe to collapse on itself. Later, this solution would be rejected as having an unrealistic premise since the measured mass of the universe was obviously not negligible.

A few years later in 1922, the Russian mathematician Alexander Friedman found a solution which was free of any preconditions or extraneous constants but which predicted a time-dependent, runaway world that was expanding in size and might eventually contract depending on the amount of matter in the universe. Sadly, hardly anyone read Friedman's papers in the obscure journal he chose to publish, and the young theorist was not able to defend his audacious predictions as he died of typhoid at the early age of thirty-seven.

This set the stage for the modest Belgian priest Georges Lemaître, who in 1927 produced a solution based on a model of the universe with constant mass. The solution *also* predicted an increasing radius for the space occupied by the universe. This accounted for the radial velocities of the extragalactic nebulae first detected by Vesto Slipher and later, with greater clarity and reliability, by Hubble and Humason at Mount Wilson. After de Sitter discovered Lemaître's published solution, Lemaître had to remind the well-known English astronomer Arthur Eddington of the paper he sent to England a few years earlier. Formerly Lemaître's supervisor at Cambridge University, Eddington quickly brought the work to the attention of the rest of the astronomical world.

By the time of the infamous 1933 meeting in Pasadena, even Einstein had accepted Lemaître's solution to his field equations, throwing his influential support behind the description of an expanding universe which Hubble and Humason had dramatically and meticulously confirmed in 1931. However, as is always the case with the appearance of a new theory, more questions developed. In particular, two quite fundamental uncertainties were brought forward: At what rate was the universe expanding? And secondly, when did the expansion of the universe begin?

Obviously, these two questions are related: the rate of expansion had already caused a crisis named by the community of astronomers as the 'timescale difficulty'. The second question, which required looking back in time over billions of years to a smaller universe and a beginning, seemed nothing short of a nightmare for cosmologists. They had no experience with any celestial bodies other than the stars, galaxies and planets which had been observed in the last few thousand years from the Babylonians to the present. This left the field of cosmology wide open for speculation and the possibility that the models of the universe that had so far been presented could be refuted.

The year 1948 was truly a remarkable watershed year for the world of physics and astronomy, as many young intellectuals were liberated from the preoccupations of the Second World War, and finally allowed to turn to science. In physics, there was sensational news. At the historic conference on quantum electrodynamics in the Poconos Mountains in Pennsylvania that year, Richard Feynman and Julian Schwinger had presented their new ideas, which had eventually won them the Nobel Prize. In the same year, the giant synchrocyclotron was commissioned at the University of California at Berkeley. The synchrocyclotron produced the first artificial mesons, a new elementary particle believed to transmit the nuclear force. Solid-state physics was also grabbing newspaper headlines with the invention of the transistor, core of the first solid-state electronic amplifier.

Astronomy was struggling to keep up with these dazzling advances. The much-heralded 200-inch (5.08-metre) telescope that had been named after G.E. Hale was dedicated at Mount Palomar in June 1948 with Hubble, Baade, Adams and the other dignitaries in attendance. However, cosmology was deemed too abstract and philosophical to attract the young physicists who were otherwise tempted by the experimentally based fields of elementary particles and solid-state physics.

In the shadow of these breakthroughs in physics, two important new theories of the expanding universe were published, both in 1948. The first was authored by George Gamow and Ralph Alpher at George Washington University. Gamow and Alpher published a totally new version of the Big Bang based on matter and radiation. This was followed by the work of three young physicists at Cambridge University in England. Fred Hoyle, Herman Biondi and Thomas Gold presented the first papers on the modern theory of the steady state universe, one of continuous creation.

But hardly anybody noticed either of these developments in 1948. In spite of their lack of attention, for the next twenty years these two diametrically opposed descriptions of the expanding universe would initiate a raging controversy, and ignite a debate that would eventually produce a clear idea of what kind of universe we live in.

George Gamow

There are few scientists in the vast collection of characters reviewed in this book who are more interesting and entertaining than the Russian-American, George Gamow. A review of his life's journey shows a man with a bizarre sense of humour and a deep fascination with the most fundamental ideas in physics and astronomy. He would eventually generate a new theory of the expanding universe – a theory that holds pride of place in today's standard cosmological model.

George Antonovich Gamow was born on 4 March 1904 in Odessa, which today forms part of the Ukraine. The son of a school teacher, astronomy fascinated Gamow from his early school years and he patiently examined the starry sky through a little telescope given to him by his father. During 1923–29 he studied optics and cosmology at the University of Leningrad (today, St Petersburg) under Alexander Friedmann. Some idea of Gamow's precocious nature is revealed in reports of his Leningrad friendships with two other brilliant students of theoretical physics, Lev Landau and Dmitri Ivanenko. The three formed a group known as the 'Three Musketeers' and met to discuss and analyse groundbreaking papers on quantum mechanics.

An early break for Gamow, which clearly affected his future, occurred in 1926 when he attended a summer school in Göttingen University, Germany. At that time, Göttingen was the world's centre of research in the new quantum theory and attracted the best minds from all over the world. His research into the atomic nucleus provided the basis for his doctoral thesis in which he applied the newly developed quantum theory to the mysterious phenomenon of natural radioactivity. In 1928, upon receiving his PhD degree from Leningrad, he went to work at the prestigious Institute of Theoretical Physics in Copenhagen, where the director Niels Bohr became very interested in his work and offered Gamow a one-year scholarship (1929–30) from the Royal Danish Academy. While working in Copenhagen, Gamow hypothesized that the atomic nuclei could be treated as little droplets of nuclear fluid. This was called the 'liquid drop' model of the nucleus, and describes a behaviour in which neutrons and protons act like the molecules in a drop of liquid. The young American John Wheeler and Bohr later adopted this theory to explain the sensational process of nuclear fission, proposing that the spherical nucleus gets distorted into a dumb-bell shape when sufficient energy is acquired by absorption of a neutron. The nucleus then splits into two fragments and releases energy according to Einstein's formula from special relativity, $E=mc^2$. These discoveries led to today's theory of fission and fusion, notorious in their application to create the first Uranium bomb and the later Hydrogen bomb.

During 1929–30, Gamow worked as a Rockefeller Fellow in Cambridge University with another leader of atomic physics, Ernest Rutherford. He immediately set to work on radioactive materials, which are known to have exponential decay rates as well as radiation emissions with certain characteristic energies. By 1928 Gamow had solved this, another important problem in nuclear physics, using the quantum theory of the decay of a nucleus via tunnelling. He used an unusual model potential for the nucleus and derived from first principles a relationship between the half-life of the particle and the energy of the emission, which had been previously discovered only empirically.

It is hard to imagine how anyone could have received a better training for the new field of nuclear physics than the young and ambitious Gamow did during the 1920s. However, in 1931, he was told to return to the Soviet Union to become a Master of Research at the Academy of Science in Leningrad. In those days Joseph Stalin was in power and though Gamow would have certainly become part of the scientific elite of the Soviet Union, the Stalin regime dismayed him. In desperation, he and his wife Lyubov Vokhminzeva, who was also a scientist, decided to leave the USSR. In their first attempt to escape, they planned to go to Turkey by crossing the Black Sea in a kayak, a journey of 168 miles (270 kilometres). After pursuing their journey for thirty-six hours they had to abandon it because of bad weather. Upon returning, Gamow managed to convince the authorities that he and his wife were carrying out some experiments in the kayak. After a few more unsuccessful attempts, in 1933 Gamow was permitted to attend the Solvay Congress in Brussels and his wife was also allowed to travel with him as his secretary. They did not return to the Soviet Union.

After receiving an invitation to lecture at the University of Michigan, Gamow and his wife left Europe for the US in 1934 and he was immediately offered a professorship at the George Washington University in Washington, DC. Before accepting the offer, he put forward three conditions. Firstly, that the University also appoint a colleague of his choice to work with him in the physics department (his choice was Edward Teller, later to become notorious for his work and promotion of the Hydrogen bomb); Second, the president of George Washington University (GWU) and the director of the accelerator laboratory at the nearby Carnegie Institution be required to organize a conference on theoretical physics to be held annually in Washington. His final stipulation was that his initial appointment at the GWU be described as Visiting Professor. Anxious to obtain a man of Gamow's ability on its faculty, the university authorities accepted all of the conditions.

In his early years at GWU, Gamow's collaboration with Teller resulted in the important theory of beta decay from the nucleus. His first work related to cosmology, the theory of the internal structure of red giant stars based on nuclear processes. During these years, Gamow took an interest in almost all branches of physics and astronomy and contributed to many. He was particularly interested in the process by which the relative abundance of the different elements in the universe might be explained by means of a naturally-occurring nuclear process such as that that occurs with stars.

The possibility of explaining the building up of atoms from elementary particles took a big leap forward when the neutron was discovered in 1932. Before that time, the atom was known to consist of a positively charged nucleus surrounded by enough negatively charged electrons to make the atom electrically neutral. Most of the atom was otherwise considered to be empty space, with its mass concentrated in a tiny nucleus. The nucleus was thought to contain both protons and electrons because the proton was the lightest known nucleus and electrons were emitted by the nucleus in beta decay. In addition to the beta particles (i.e. electrons), certain radioactive nuclei emitted positively charged alpha particles and neutral gamma radiation.

Lord Ernest Rutherford, a pioneer in atomic structure, had postulated the existence of a neutral particle with the approximate mass of a proton that could result from the capture of an electron by a proton. This stimulated a search for the particle. However, its electrical neutrality complicated the search because almost all experimental techniques of this period could only measure charged particles. Finally, James Chadwick at the

Cavendish Laboratory in Cambridge proved that the emissions he observed from a scattering experiment contained a neutral component with a mass approximately equal to that of the proton. He called it the 'neutron' in a paper published on 17 February 1932. In 1935, Chadwick received the Nobel Prize in physics for this work. The availability of a free neutral particle with the mass of a proton dramatically opened up the field of nuclear physics. In 1934, while at the University of Rome, the Italian physicist Enrico Fermi began experiments bombarding all the elements in the periodic table with neutrons. New elements seemed to be produced and it was thought that neutrons were being captured by the nuclei of the bombarded elements.

It soon occurred to Gamow that neutron capture might be the basic mechanism in the formation of heavier nuclei in stars. Though this was not an original idea to Gamow, but as a nuclear physicist par excellence, he was quite capable of understanding all the details. He stated this in a lecture at Ohio State University in 1935 in which he also entertained his audience by paraphrasing an Arthur Eddington remark from the past: 'What might be possible in the laboratory of Enrico Fermi in Rome might not be too difficult in the interior of the Sun.'

Gamow's work referred to neutrons that could be ejected from the nuclei of light elements by collisions with protons and then stick to the nuclei of different heavy elements, thus causing the formation of still heavier nuclei. This could take place in a highly energetic atmosphere such as the hot interior of the Sun. Gamow based his idea on the process by which a neutron would react with a nucleus of mass number A and atomic number Z, producing a beta radioactive isotope (by emitting an electron). As a result of the decay, the isotope would transform into a nucleus with unchanged mass number but with atomic number increased by 1, that is, an element with a higher position in the periodic table, just as Fermi's experiments anticipated.

Thus it was only from about 1932, with the discovery of the neutron, that nuclear physics could contribute fruitfully to the understanding of cosmic phenomena, in this case, processes related to attempts in understanding the energy production in stars like the Sun. Yet, as far back as 1920 when Arthur Eddington's Cambridge colleague, Francis Aston, was making precise measurements of atomic masses, Eddington noted that four hydrogen atoms have slightly more mass than one atom of helium. He suggested that if Helium could be produced from combining Hydrogen, the energy equivalence of this mass difference, from Einstein's $E=mc^2$ equation, could easily power a star's energy needs.

Hans Bethe: How the Sun Shines

The next giant step in understanding the production of solar energy was taken by the German-American nuclear physicist Hans Bethe, who turned up in 1938 at one of Gamow's Washington conferences. He came away with some big ideas. Bethe had studied under the great teacher Arnold Sommerfeld in Munich where he obtained his PhD in 1928 and then held positions at various German universities. He was quickly seen to be one of the bright young theoretical physicists of his generation and excelled in quantum theory, which was so important in Germany in the 1930s. However, because he was half Jewish he fled Germany in 1933 and settled at Cornell University in Ithaca, New York. There he rapidly established himself as a leading authority in nuclear theory.

At the 1938 conference at GWU, Bethe saw the possibility of solving the problem of explaining the energy of the sun by applying nuclear physics. He began work immediately on the problem and started to get some initial results even before the conference was over. As legend has it, Bethe worked out the main outline of the solution while travelling by train from Washington back to Ithaca. Less than half a year later he had produced a detailed calculation of the reactions of protons with carbon and nitrogen nuclei. He later called this the carbon-nitrogen (CN) cycle, and was awarded the Nobel Prize in physics for this groundbreaking work. William Fowler, who would years later win a Nobel Prize for showing how the elements were cooked in stars, recalled Bethe's influence:

> Bethe's paper of 1939 told us that [what] we were studying in the laboratory were processes which are occurring in the sun and other stars. It made a lasting impression on us. Bethe was thus arguing that the carbon-nitrogen cycle was the principal energy source for stars on the main sequence of the H-R diagram, which includes the Sun. With this work, stellar energy was turned into a laboratory science. After the war, Bethe's ideas would be extended to cosmology as Gamow contemplated the early universe.

Gamow was a prodigious communicator, always keeping in touch with the best contributors in his field of interest. He corresponded prodigiously, supervised students, organized conferences and travelled everywhere to deepen his understanding of nuclear physics as applied to major problems in cosmology. His experience in various aspects of atomic physics had uniquely prepared him to investigate Bethe's work

The first important date of what one might call *nuclear cosmology* was in 1938, when the German, Carl F. von Weizsacker proposed his theory of stellar energy production and cosmic nucleosynthesis, which can be defined as the building up of the elements in stars and possibly elsewhere in the cosmos. His theory was soon overshadowed by Bethe's much more detailed theory, but from a cosmological point of view, Weizsacker's work was far more interesting. This is because he suggested, albeit in a preliminary way, the idea that matter in the universe today is the result of nuclear processes located either in stars or in other hot, dense environments in the distant past before stars had formed. This presented the possibility that by comparing the result of nuclear-physical calculations with the observed cosmic distribution of chemical elements in the universe today, some insight into the physical conditions of the early universe could be realized.

In 1938, Weizsacker proposed that a nuclear physicist could study the nature of the distant past of the cosmos by studying the chemical composition of the universe. It was an idea that was intriguing to Gamow. However, a research programme like this could only hope to succeed if the relative abundance of nuclei in the universe was known on a cosmic scale. Coincidentally, such knowledge appeared in the same year, 1938, primarily as a result of the painstaking analyses made through more than a decade of research by geochemists, astronomers and meteorite specialists. This did not go unnoticed by Gamow, who was always looking for projects that would test his profound knowledge and experience with nuclear physics. Since the final answer was known, i.e. the relative abundance of the elements in the universe, Gamow was interested in the possibility of speculating on the interactions and nuclear processes which could produce these relative abundances.

Though Gamow's main interest, like Bethe's, was in nuclear reactions inside stars, there are some indications that he was also considering the nature and mechanisms of the early universe. In a review article from the summer of 1938, he wrote: 'Since the physical conditions of an epoch so long ago are highly hypothetical, a broad area for speculation about the origin of stars is opened up.'

A year later, he pursued the possibility that the early universe may have formed all the elements in a 'nuclear oven'. In a popular book, *The Birth and Death of the Sun*, produced that year, he wrote; 'A long, long time ago the universe consisted of a primordial gas of extreme density and temperature that gradually decreased [i.e. in temperature] as a result of expansion.'

Clearly influenced by Weizsacker's ideas that the heaviest elements must have been produced under extreme conditions of temperature and pressure, Gamow was now considering the possibility of a super dense and super hot early universe. It appears that this was the first time he seriously entertained the idea of the Big Bang universe.

Though Gamow was primarily a nuclear physicist and not an astronomer, he did have some experience with cosmology. He was, after all, a former student of Alexander Friedmann and he also wrote papers with Teller, his colleague at George Washington University, on the gravitational compression of matter into galaxies. It may be more appropriate to call him an astrophysicist. Around 1940 it was not uncommon for astrophysicists to refer to the possibility that the chemical elements may have formed in a much earlier state of the universe where matter was compressed to nuclear densities. It was also not uncommon to connect this hypothetical state with the pre-expansion of the universe as envisioned by Lemaître, to align the concept of nuclear densities with his primeval atom.

However, there seemed to be no great advantage in speculating that the elements were formed cosmologically when the possibility that they could be formed in the interior of stars was another, much less speculative option. It was only later when it became apparent that the stars were unable to produce a wide range of elements in the right ratios that some astrophysicists found it necessary to focus on the option of pre-stellar formation. This opened up the possibility of a Big Bang universe.

The Big Bang theory was also the conclusion reached by Weizsacker and Bethe, who seemed to suggest that the elements had had a non-stellar origin, perhaps in an earlier state of the universe. Gamow was acutely aware of these developments, and they soon enticed him into cosmology, an example of an important scientific problem involving the nucleus, which he could not resist.

The last of the Washington conferences, which occurred in 1942, was important for Gamow and the Big Bang. At this meeting the participants agreed that a hot, dense, early universe was necessary to account for the existence of heavy elements. The final conference report written by Gamow and a colleague from the famed Carnegie Institution concluded that:

It seems plausible that the elements originated in a process of an explosive character that took place at the beginning of time and resulted in the present expansion of the universe.

Shortly after this meeting Gamow tried to promote a physical picture of the beginning of the universe as being similar to the well-known fission process that resulted in the Uranium bomb, but this never caught on. Nevertheless, busy with many different projects during the war years as well as his academic responsibilities at George Washington University, he continued to sketch out ideas, postulating the ways in which the elements might be formed by the elementary particles of the nucleus under the conditions of the early universe. A breakthrough in his research came just after the war ended when Gamow wrote to Niels Bohr on his sixtieth birthday in October 1945:

> ... what I'm trying to do at present is study the problem of the origin of elements at the early stages of the expanding universe. It means bringing together the relativistic formulae for expansion and the rates of thermonuclear and fission reactions. One interesting point is that the period of time during which the original fission took place (as estimated by the relativistic expansion formulae) must have been less than one millisecond, whereas only about one tenth of a second was available to establish a subsequent thermodynamic equilibrium (if any) between the different lighter nuclei. I am planning to have our next conference here in the spring on that problem and the other problems of the borderline between nuclear physics and cosmology.

Thus in the fall of 1945 Gamow conceived the essential idea of an explosive model based on combining nuclear physics at the beginning of the universe with the Friedman-Lemaître equations for the expansion. At this stage he still believed the nuclear process involved was a *fission* reaction, but this was soon to change.

In September 1946 he sent a brief paper to the *Physical Review* which is often hailed as laying the foundation of modern Big Bang cosmology. Based on Hubble's convincing evidence in support of the expansion of the universe, this theory was a different type of 'Big Bang' from Lemaître's mathematical expansion theory because it was motivated by consideration of the initial formation of the chemical elements. In the paper, Gamow combined two perspectives, neither of which was new, but taken together served as a foundation for much of the further development of the model.

To begin with, he repeated the conclusions of the Washington conference of 1942, and stipulated that the chemical elements must have been formed during an explosion that took place at the beginning of time and that this resulted in the expansion of the universe. He then made a decisive step to connect the explosion with the expansion described by Lemaître's relativistic theory based on a special solution of Einstein's field equations.

Gamow immediately realized that the rapid expansion of the universe made it impossible to consider the early formation of the elements as a process of equilibrium. Instead he imagined the earliest universe to consist of a gaseous soup of neutrons. This picture, a kind of very large atomic nucleus made up only of neutrons, was not unlike the primeval atom that Lemaître proposed in 1931. Yet it is important to understand that Gamow was setting out a new theory of the big bang – a *physical* cosmology based on the earlier *mathematical* cosmology – as a part of the model.

By 1946 Gamow was ready to seriously develop his explosive-nuclear model of the early universe. In particular he wanted to see if the model predicted the known abundances of the chemical elements. He needed some help with the complicated calculations and decided to take on the supervision of a PhD graduate student. He would assign the *primordial nucleosynthesis*, or the formation of the chemical elements in the early universe, as the thesis topic. He was fortunate to interest Ralph Alpher, a bright hard-working young local physicist, the twenty-five year old son of émigré Russian Jews. Gamow had already known Alpher for a few years and was partial to him because Alpher's father was also from his birthplace, Odessa.

Alpher was a somewhat unusual graduate student, mature and fiercely determined to succeed. In the early 1940s, he had started attending night classes at GWU given by Gamow. There he shared Gamow's fascination with the origin of the elements that make up the world. At that stage there was increasing evidence that the universe had started out with only atoms of the simplest element, hydrogen, and that all others had been assembled, step-by-step, by the addition of basic atomic particles. Yet where, when and how did this happen? It had to be somewhere intensely hot, a nuclear oven, because ionized hydrogen atoms normally repel each other. This repulsion can be overcome only if they collide at high speed, which in practice means at ultra-high temperature. The stars were extremely hot but these particles would have to combine even before the stars formed. Gamow, and several others at the Washington conference of 1942, had decided the early universe, not the stars, provided the furnace for the elements to forge.

Gamow, who quickly became bored with laborious calculations, gave Alpher a huge challenge to calculate how neutrons could be captured by protons i.e. hydrogen atoms. To transform the idea into a quantitative theory, experimental knowledge of the nuclear reactions that are involved was needed. It was therefore fortunate that precise data on the cross-sections of fast neutrons captured by wide range of elements had just appeared. The data had been obtained at the Argonne National Laboratory in the US. Alpher had heard a ten-minute talk at a meeting of the American Physical Society in which these data were mentioned, and immediately requested a full set of data from Argonne.

Hard-working and careful by nature, Alpher started crunching numbers and made an amazing discovery. He found a narrow window between about 30 seconds and 300 seconds after the big bang during which the universe was hot and dense enough for element-building to occur. He was able to predict the present levels of hydrogen and helium in the universe (which make up over 99 per cent of all matter) using a progression of nuclear reactions that occurred just at the beginning of the expansion of the universe. Alpher worked out how the temperature and other properties of the universe changed as it expanded. He also continuously adjusted the values of nuclear reactions, inserting them into the time-varying environment to get the desired result – the correct proportions of hydrogen, helium and the heavier elements in the universe, past and present.

Gamow was impressed with the work of his assistant, and even before Alpher's PhD dissertation was completed, he convinced the young graduate student that they should publish their results immediately. 'The Origin of Chemical Elements' was thus submitted to the *Physical Review* and published in April 1948. The paper is considered one of the most important on the Big Bang theory of the universe. Though built mainly on an incomplete dissertation, and incorrect in many important details regarding the heavy elements, the paper did give the correct relative abundances from a new picture of the earliest universe and indicated the approach to be followed in future research.

From the point of this publication onwards, the very early universe was now described as a hot, highly compressed neutron gas which at some time started decaying into protons and electrons, forming at least the first four chemical elements in their correct proportion in the universe. The origin of the neutron gas was left as an initial starting condition without explanation.

The paper which was signed by Alpher, Gamow and strangely, Hans Bethe, is often referred to as the 'alphabetical article' and has become famous, or perhaps infamous for this reason. Gamow, always the joker, humorously decided to add the name of his friend the eminent physicist Bethe in order to create the whimsical author list of Alpher, Bethe & Gamow, forming a pun on the first three letters of the Greek alphabet, αβζ. Bethe was listed in the article as H. Bethe, Cornell University, Ithaca, New York (in absentia).'

Gamow explained Hans Bethe's association with the theory in this way:

In writing about the preliminary communication of this work, I was unhappy that the letter beta was missing between alpha and gamma. Thus, sending the manuscript for publication in the *Physical Review*, I put the name of Hans Bethe (in absentia) between our names. This was planned as a surprise to Hans when he would unexpectedly find his name as co-author. I was sure that being my old friend and having a good sense of humor, he would not mind. What I did not know was that at the time he was one of the reviewers for the *Physical Review* and that the manuscript was sent to him for evaluation. But he did not make any changes to it except to strike out the words (in absentia) after his name, thus endorsing the idea and the results.

Alpher, at the time only a graduate student and one of Gamow's acolytes, was generally irritated by the addition of Bethe's name to this paper. He felt that the inclusion of another eminent physicist would overshadow his personal – and significant – contribution to this work and prevent him from receiving proper recognition for such an important discovery. He expressed resentment over this as late as 1999, years after Gamow's death.

In the spring of 1948, Alpher and Gamow realized the early universe was dominated by radiation, not matter, with a temperature of about a billion degrees Kelvin. In subsequent papers, they wrote of a 'radiation filled universe'. But as the universe expanded, they wrote, the domination of radiation over matter would decrease until at a certain time, the two densities would be equal. As the universe continued to cool to a thousand degrees Kelvin, matter would begin to clump together and galaxies would form. Although the numerical estimates were wrong, this was the first time a 'crossover' idea which expressed the decoupling between radiation and matter, was ever discussed.

Gamow was also convinced that the residual radiation left-over from the initial super dense state was still somewhere in the universe, as it had no place to go. In a way, the fireball of the Big Bang is like the fireball of a nuclear explosion. Whereas the heat of a nuclear fireball eventually dissipates into its surroundings, the Big Bang fireball must still be around. Of course, the expansion of the universe over the past 13.7 billion years has diluted and cooled it. As early as 1948, Gamow used estimates of the cosmic density of particles to predict, for the first time, the existence of a radiation background left over from the early universe. He calculated that by the present time the Big Bang fireball would have cooled way down to a temperature of only five degrees Kelvin, hardly a 'fireball' anymore. A few years later, in his popular book *The Creation of the Universe*, this estimate increased to fifty degrees Kelvin.

Though in the original paper no estimate of the residual cosmic background radiation was made, Alpher focused on this aspect of the theory shortly after the publication when he started working with Robert Herman whom he had met at the Johns Hopkins Laboratory. Herman's family background was similar to Alpher's. His parents were Russian Jews who arrived in New York City in 1910, where Robert was born four years later. After undergraduate studies at City College New York, he received his PhD in physics at Princeton University where he had become acquainted with relativity and cosmology. Contrary to the younger Alpher, Herman had more experience and had already published many research papers. He would replace Gamow as Alpher's coworker on the details of the Big Bang theory.

Alpher and Herman published their first results on the Big Bang theory in the scientific journal *Nature* in 1948. They calculated that the fireball radiation should no longer appear in the form of visible light but in the form of microwaves invisible to the unaided human eye. In a series of papers extending over a five-year period, they gradually refined the calculations in the original Alpher-Bethe-Gamow paper and got promising results with regard to the lightest elements, but were unable to suggest that nuclear reactions had produced the heavier elements at the Big Bang.

Gamow's approach to early Big Bang cosmology culminated in a masterful paper published in 1953 by Alpher, Herman and another colleague from the Johns Hopkins Laboratory, James Follin. These three physicists made detailed calculations for the first ten minutes after the Big Bang using all available know-ledge of current particle physics and the most recent advances in nuclear and particle theory. This was, in fact, the first thoroughly modern analysis of the early history of the universe, and marked the climax of the Gamow-Alpher-Herman Big Bang approach. With this report, the theory was brought to a state that would not change much over the next fifteen years. What is more, despite the fact that the Johns Hopkins alumni repeated the 1948 prediction of the background radiation seven times in their published literature, the idea had no impact.

Many science historians have speculated on the reasons why astronomers did not make any effort in those years to detect the background radiation. There was a complete lack of interest in the prediction and Alpher and Herman were frustrated as they were not able to make the measurements themselves. They repeatedly asked radio astronomers whether the fireball radiation was detectable and were repeatedly told – incorrectly – that it wasn't. Gamow, adept at making concepts popular, sensationalized the 'afterglow of creation' in magazines and books, but still nobody took up the challenge. Why were astronomers so reluctant to attempt to detect the background radiation? To answer this question, a basic understanding of the electromagnetic spectrum is necessary.

Recall that Gamow's original prediction about the early universe described a hot, dense, rapidly evolving plasma (called 'ylem' by Gamow and his associates) that was a mixture of matter and radiation. In his original paper, he was able to make some estimates of the temperature at which the matter and radiation would separate, or decouple. Alpher and Herman later refined this estimate to the value of about 300,00°K. To further study radiation, researchers had access to a well-defined theory describing the radiation given off by any body that has a temperature. This was called the theory of thermal radiation.

The Thermal Radiation Spectrum

When a body is heated, it emits radiation over the entire spectrum of electromagnetic waves. The distribution of the emission is a function of the temperature as shown in the diagram for 800° C, 300°C and just 5° absolute (Kelvin) temperature.

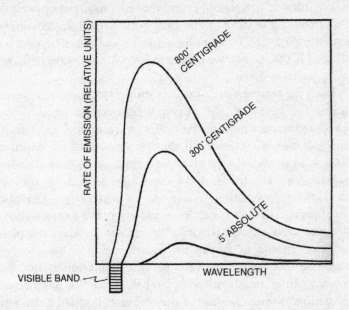

Thermal radiation intensity vs wavelength at different temperatures.

The question is how to measure this radiation, or which wavelength band to search. The confirmation of the existence of the cosmic background radiation would be one of the most important discoveries in the history of science. But could it actually be verified?

The underlying physics of thermal radiation is quite simple, although it did require the radical hypothesis that Max Planck put forth in 1900 to explain the details. His theory showed how the relative rate of the emission of radiant energy (electromagnetic waves) from a hot body at a certain temperature is different for different wavelengths. Planck's theoretical curves show that radiation spreads out and that the peak shifts to longer wavelength as the temperature drops. At 800°C, enough visible radiation is emitted to appear red hot, though most of the energy emitted is in the infrared band. At 300°C, practically all of the energy emitted is carried by waves longer than red light and are called 'infrared', meaning beyond the red. No radiation is emitted in the visible band at this temperature. At 5 degrees above absolute zero -268°C, all radiation being emitted is beyond the infrared in the microwave band and special microwave receivers are required to make measurements.

Since the temperature of the emitting body uniquely determines the shape of the curve, measurements at different wavelengths can indicate the temperature of the body emitting radiation. Conversely, if the temperature of the emitting body is known, then the shape and the distribution of radiation can be predicted from the theoretical formula. Returning to Gamow's prediction (as refined by Alpher and Herman) the theoretical curve for thermal radiation at 5 degrees above absolute zero indicated that in the universe today, the peak radiation should be in the microwave region of the electromagnetic spectrum. This was a result of the thinning out and cooling of the initial white-hot fireball of the Big Bang under the action of the expansion of the universe. Radiation was still present, but wavelengths have been stretched all the way to the microwave band by the expansion.

This was unfortunate for Gamow and his associates because in the 1950s, observation of radiation in the microwave band was impractical and there were only a few observatories in the world where such measurements could be made. This was due to the fact that entirely different devices were required. The giant telescopes like Mount Wilson and Palomar in California were useless for this measurement. Certainly, this was one drawback to the proposed experiment, but there were other factors that kept radio astronomers from being enthusiastic about the predicted background radiation. Nobel Laureate nuclear physicist Steven Weinberg spelled these out in the popular book, *The First Three Minutes*. He explained why the importance of a search for a 5-degree Kelvin microwave radiation background was not generally appreciated in the 1950s and early 1960s.

In addition, the physical evidence was also missing. Early Big Bang theorists were mostly nuclear physicists, yet they could not demonstrate how elements with a higher mass than helium could be produced in the early universe. Physicists soon noticed that these 'mass gaps' would hinder the production of all the higher elements. Just as it's impossible to climb a staircase one step at a time when one of the steps is missing, this discovery meant that the successive neutron capture theory could not account for higher elements in the universe.

Gamow, ever optimistic about the theory he had spent so many years promoting, believed that this shortcoming should not hamper the acceptance of the entire Big Bang model. The theory had accounted for the formation of all the observable hydrogen and helium, 99.99 per cent of all the matter in the universe. Although he acknowledged that he had not uncovered how the heavy elements were formed by the Big Bang, he believed the problem was not *insolvable*, just merely *unsolved*. And, as it turned out, he was right.

British astrophysicist Fred Hoyle, who later became a bitter critic of the Big Bang model, demonstrated that the heavier elements had not necessarily been produced in the early universe by the Big Bang. In a seminal forty-page paper presented to the Royal Astronomical Society in London in November 1946, he was able to show how the synthesis of the elements could proceed from hydrogen in nuclear reactions occurring during the evolution of stars in our galaxy thousands and even billions of years after the Big Bang. These elements could then spread throughout the galaxy by the explosions of supernovae. Hoyle had discovered a key reaction in the stellar sequence which would take place at the necessary rate only if Carbon-12 had a very particular property, technically, an energy level just above the critical energy for combining beryllium and helium nuclei. Hoyle predicted this energy level and his colleague Willy Fowler confirmed its existence experimentally. One of Hoyle's colleagues, Margaret Burbridge described the atmosphere of that day in London when Hoyle first made the prediction:

> I have never forgotten the experience of listening to Fred give this exciting account of his work on building the elements in the abundance peak where statistical equilibrium would prevail in high temperature high density interiors of the stars.

Begrudgingly, Hoyle later had to admit that his calculation was also consistent with a Big Bang cosmology: 'any super dense state of matter that may occur in any evolutionary cosmology satisfies the requirement that matter emerges from the super dense state essentially as hydrogen.'

In the end, Gamow was grateful to Hoyle for this contribution on the mass gap problem, inspiring his wife to pen a biblical parody placing Hoyle in the role of an assistant in the creation of the world:

The New Genesis
In the beginning God created radiation and ylem. And ylem was without shape or number, and the nucleons were rushing madly over the face of the deep. And God said: 'Let there be mass two.' And there was mass two. And God saw deuterium, and it was good. And God said: 'Let there be mass three.' And there was mass three. And God saw tritium and tralphium [Gamow's nickname for the helium isotope He-3], and they were good. And God continued to call number after number until He came to transuranium elements. But when He looked back on his work He found that it was not good. In the excitement of counting, He missed calling for mass five and so, naturally, no heavier elements could have been formed. God was very much disappointed, and wanted first to contract the universe again, and to start all over from the beginning. But it would be much too simple. Thus, being almighty, God decided to correct His mistake in a most impossible way. And God said: 'Let there be

Hoyle.' And there was Hoyle. And God looked at Hoyle . . . and told him to make heavy elements in any way he pleased.

And (to continue the parody) Hoyle decided to make heavy elements in stars, and to spread them around by supernovae explosions. But in so doing he had to obtain the same abundance curve that would have resulted from nucleosynthesis in ylem, if God would not have forgotten to call for mass five.

And so, with the help of God, Hoyle made heavy elements in this way, but it was so complicated that nowadays neither Hoyle, nor God, nor anybody else can figure out exactly how it was done.

Fred Hoyle spent most of his entire career at Cambridge University in England. His theories, like Gamow's, were not always correct, yet his importance in the history of cosmology is immense. He championed a radical new theory – the steady state theory – of the universe based on the continuous creation of matter, which in the end, turned out to be wrong. However, his harsh criticism of Gamow's Big Bang alternative would invite intense scrutiny of that model which eventually – and ironically – convinced scientists that it was, in fact, correct. This process was controversial and acrimonious, lasting over twenty years from 1948, the year both theories were introduced.

Hoyle was undoubtedly one of the most colourful scientists of the twentieth century. Described by most of his contemporaries as a maverick with a clear dislike for orthodoxy, he possessed an unusually abrasive personal style and a penchant for speaking his mind. This was a breath of fresh air in the great halls of academia, but Hoyle carried it too far. He relished disagreeing with just about everyone, even his co-workers, and could never admit to being wrong. He thrived on controversy, in his own discipline and beyond, and nourished an attitude which would ultimately poison his exceptional career, ruin his chance for a Nobel Prize and cause him to turn his back on Cambridge University where he had risen to a position of great influence as the first director of the Cambridge Institute of Theoretical Astronomy.

One could say that Hoyle entered university life with a chip on the shoulder, arriving in Cambridge as a complete outsider with a pronounced Yorkshire accent, the son of a father with no job or profession, wearing thick walking boots and a shabby academic gown. In spite of the disadvantages that he was faced with, Hoyle was always able to gather support for his education. The head of his local grammar school, a trained mathematician, helped him get into Cambridge as the first of many fortunate steps placing him in the right place at the right time.

In the 1930s at Cambridge, he learned quantum mechanics from Paul Dirac, the sublime mathematician who contributed substantially to its theory, later winning the Nobel Prize for his work. He learned general relativity from his hero, Arthur Eddington, who was credited by Albert Einstein as writing more clearly about that new theory of gravitation than anyone else. These legendary physicists taught Hoyle not only the inside story of their revolutionary ideas, but also the critical lessons that enable any good theorist to avoid the obvious and look for paradoxes which needed to be solved.

There is no denying the importance of the many and varied aspects of Hoyle's work. He single-handedly re-established the international prestige of British astronomy in the 1950s; he carried out the complex calculation which went on to form the basis for understanding the formation of carbon (and the remainder of the chemical elements) in the interior of stars; and he was a pioneer in the popularization of modern cosmology in his writing and radio broadcasts. However, he could not temper his determined, but inflexible efforts to prove that his own steady state theory was correct in the light of evidence to the contrary.

Consequently, Hoyle is remembered for this mistake as much as any other (mostly correct) contribution he made in hundreds of technical papers and books. Oddly, it was not the

American originators of the Big Bang theory who first attacked the steady state model in the 1950s and 1960s, but Hoyle's colleagues at Cambridge. This provoked a cause célèbre on the campus that ruined the later portion of Hoyle's career.

Hoyle first met his steady state collaborators, Austrian Jewish refugees Hermann Biondi and Thomas Gold, as a young man during the Second World War when he worked on radar applications for the Admiralty Signals Establishment. In between assignments to enhance the British war effort, the three men spent their spare time on scientific discussions that eventually led to the development of the theory. Biondi credits Hoyle with inspiring him and Gold in these early days to take up the big questions in cosmology.

> . . . When Fred stayed over [at Biondi's farm cottage] we spent all of our time discussing scientific problems. Fred's enormously stimulating mind, his deep physical intuition, his knowledge of the most interesting problems in astronomy, all combined to give me an outstanding scientific education in the few hours left after a hard day's work . . . Fred Hoyle is always full of ideas in astrophysics, full of problems with which neither Tommy (Gold) nor I were acquainted. When he stayed with us we talked late into the night about these questions.

Discussions continued between the three when they all arrived at Cambridge after the war. In 1948, their revolutionary papers on the steady state theory were published and although the names of Biondi, Gold and Hoyle have always been linked together in reference to the theory, Hoyle's contribution was actually published separately. This is because Hoyle wanted to distance himself somewhat from the purely philosophical ideas of his colleagues. He submitted his paper to both *Nature* and the *Physical Society*, both of which rejected the work. He therefore had to settle for publishing his paper with the Royal Astronomical Society, where Biondi and Gold had published two months earlier.

The philosophical character of Biondi and Gold's hypothesis was based on 'the perfect cosmological principle' in which the universe is said to be unchanging not only in space but also in time, thus escaping the nasty problem associated with establishing a beginning in time. Since the universe was known to be expanding (no one questioned Hubble's observations on Mount Wilson), how could the universe always look the same?

This is where Hoyle's contribution to the theory was important. According to Hoyle, matter was continuously being created and new stars and galaxies are formed to fill the space left behind as the old ones moved apart in the expansion. He was not happy, however, to base the steady state theory solely on the 'perfect cosmological principle', and he further explained how primordial hydrogen could be continuously created to maintain the steady state. He placed the concept within the framework of general relativity. With his powerful command of mathematical physics, Hoyle believed it was straightforward to propose a modification to Einstein's relativistic equations. He therefore simply added a term for the creation of matter, which he termed the 'C' field – C for 'creation'.

Hoyle, Gold and Biondi promoted the steady state theory at conferences, in publications, in the media and to their graduate students. Yet Hoyle's alienation from the astronomy establishment had not subsided since the furore he created with remarks he broadcast on BBC Radio 3 in March 1949. During the programme in question, he announced his endorsement of the steady state model to a large popular audience. He also coined the term 'Big Bang' during the same broadcast. Historians claim Hoyle intended this tag to be pejorative, but the script from which he read clearly shows that he intended the expression to be helpful to listeners. Hoyle later explicitly denied that he was being insulting and said that the term was a striking image meant to emphasize the difference between the two theories for radio listeners. Nevertheless, a clear confrontation was building between the steady state and big bang models.

Big Bang versus Steady State

Given that the steady state and Big Bang theories were intended to explain the same universe, it is appropriate to examine how these two models differ. In particular it is worth looking at three experimental facts: the expansion of the universe as it was measured by Edwin Hubble on Mount Wilson; the measured relative abundance of the primordial elements in the universe, i.e. hydrogen and helium; and finally, the distribution of galaxies throughout the universe.

If new or 'young' galaxies are only found at great cosmological distances, then the steady state theory – which predicts the continued generation of new matter and thus a uniform distribution of young and old galaxies – clearly must be wrong. Big Bang theorists would say these young galaxies were at the front edge of the expansion of the universe. On the other hand, if the microwave background radiation that is predicted by the Big Bang theory is never detected, then this theory is flawed.

Both of the proposed theories were able to account for the first two observed properties of the universe, the expansion of the galaxies and the abundance of at least the lightest primordial elements. The third proposition, however, demanded a new step in astronomical theory. However, not until the 1950s did astronomers begin to explore the further reaches of the electromagnetic spectrum. This consisted of waves beyond the infrared with longer wavelengths than the optical or visible band. After the Second World War, these frequencies would be probed for any new information they could provide. A new generation of astronomers, called *radio astronomers*, who believed they had an advantage over optical observations in obtaining new information on the universe, began to emerge.

Jodrell Bank and Radio Astronomy

In England, the advent of radio astronomy was a direct outgrowth of war-time research on radar. Jodrell Bank at Manchester, the first British radio observatory, was established in 1945 by Sir Bernard Lovell, who wanted to detect faint signals from outer space and investigate cosmic rays. This pioneering institution has since played an important role in the research of meteors, quasars, pulsars, masers and gravitational lenses, and was heavily involved with the tracking of space probes at the start of the Space Age.

At Cambridge, the radio astronomy effort was led by Martin Ryle, who had done salutary work in the radio band during the war, jamming and diverting German communication systems. After the war, he looked for applications of radio waves and was attracted to cosmological questions. At first he believed that all sources of radio waves from space were located within the Milky Way galaxy and hence were of no cosmological importance. But over the next few years, as he continued to develop his ambitious measurement programme, Ryle became convinced that most of the radio sources he was detecting were, in fact, extragalactic, that is, existing outside of the Milky Way galaxy.

His observations could thus be used to test cosmological models and threaten the smug confidence of the theoretical cosmologists, whose brief he believed was over-extended:

> Even if we never actually succeed in [radio] measurements with sufficient accuracy to disprove any cosmological theory, the threat may discourage too great a sense of irresponsibility [on the part of cosmologists, who] . . . have always lived in a happy state of being able to postulate theories which had no chance of being disproved.

Ryle was particularly critical of the steady state theory, the brainchild of his colleague at Cambridge, Fred Hoyle. For his part, Hoyle complained that Ryle was motivated not by a quest for the truth, but by a desire to destroy the theory. Simon Mitton, who worked with both men at Cambridge, has written a superb biography of Hoyle, and set the stage for the Ryle-Hoyle confrontation by describing the background and personality of both. This is what he had to say about Ryle:

Ryle, aged twenty-three years became group leader (of the Cavendish Laboratory at Cambridge), taking charge of electronic countermeasures and deception . . .

As head of the group, Ryle's extraordinary inventiveness, his scientific understanding, his management skills and his capacity for sheer hard work came to the fore. But the atmosphere was full of desperate anxiety. He was a tall individual who could be aloof. Under the stresses of urgent operational requirements, he was intolerant of those who did not share his immediate insight. According to a senior colleague, Ryle could be 'highly temperamental and not an easy man to work with'. There is no doubt he would speak loudly, clearly and passionately when provoked . . . This character trait gave an angry sharpness to his disagreements with Hoyle.

After the war, Ryle returned to Cavendish, deeply affected by the tragedies of war, notably the terrible effects of British air raids on German civilians. Having worked on countermeasures to protect British bombers, he felt deep personal responsibility and this affected his outlook on his future career. Ryle wrote:

> By the end of the war we were all very tired. Few of us knew precisely what we wanted to do. I was very tense and . . . I certainly knew what I didn't want to do. I wanted nothing more to do with military equipment. I was not one of those who would be content designing bigger and better radars in preparation for the next war.

Making good on his promise to develop peacetime applications of his radar experience in radio astronomy, Ryle and his team carried out systematic surveys of almost 2,000 radio sources. Though the first survey was sketchy and his interpretations of the results shown to be inaccurate, the second survey, completed in 1955, seemed to contradict the predictions of the steady state theory. More distant radio sources seemed to be distributed differently from those nearby. Though most historians of this period criticized Ryle for overstating the significance of his initial data, after a few more years of work his observations would support arguments that were increasingly pitted against the steady state theory. Naturally, Hoyle and other steady state enthusiasts pointed to the earlier discredited results and emphasized that radio astronomy data was unreliable. As far as the three developers of the theory were concerned, the steady state model was still alive and well. This controversy set the stage for a most remarkable development.

The Most Important Accidental Discovery in History

It appears that what is arguably the most important discovery in the history of science was made completely by accident. In fact, the two scientists who made the discovery had no idea what they had uncovered, even though twenty years later it would win them the Nobel Prize for Physics.

The story begins in America's two top institutions for basic research in physics: Princeton University and the Bell Telephone Laboratories, both in New Jersey and less than an hour's drive apart. At Princeton astrophysicists Robert H. Dicke, Jim Peebles and David Wilkinson were exploring the theory of the early universe, taking a fresh look at the theoretical consequences of the Big Bang. They quickly reasoned that not only matter was scattered at the beginning and later condensed into galaxies, but that radiation was also released in a tremendous flash of energy. They concluded that the radiation still pervaded the universe, and that with proper instrumentation it could be detected.

The Princeton group was therefore preparing to search for the Big Bang radiation in the microwave region of the spectrum, just as Alpher and Herman had hoped fifteen years earlier. The leader of the group, Dicke, had distinguished himself in his work in the Radiation Laboratory at the Massachusetts Institute of Technology (MIT) during the Second World War. Born in St Louis, Missouri, Dicke completed his BA at Princeton University and his doctorate, in 1939, from the University of Rochester in nuclear physics. While working on the development of radar in the 1940s at MIT, he invented the microwave receiver, now called the *Dicke Radiometer*. This device was capable of detecting low levels of radiation in the microwave band. During the 1960s, this device was used by radio astronomers everywhere, including Bell Laboratories, for tracking their early communication satellites Echo-1 and Telstar.

A rare combination of experimentalist and theorist, Dicke began to think about the early universe after arriving at Princeton in the early 1960s. With Jim Peebles, he re-derived the prediction of a cosmic microwave background. Dicke had been working on different aspects of cosmology and had even attempted an alternative version of Einstein's general relativity. These studies resulted in his interest in the Big Bang. With David Todd Wilkinson and Peter G. Roll, Dicke immediately set about building a radiometer that could search for the background radiation.

Unlike Gamow and his coworkers, the Princeton team had in-house experience in radio signal measurements when they set out to construct instrumentation to detect the background radiation. Completely oblivious to the papers that the Gamow team had published in 1948 and 1952, the Princeton group independently calculated the characteristics of the radiation, obtaining almost the same temperature and wavelength as the earlier work, in the microwave band. At that point, Peebles decided to write up the details of the calculation and send a draft to several of his contacts working in the same field. One of those who received a copy was Bernard F. Burke, Professor of Physics at MIT.

In December 1964, Burke had attended an astronomy conference in Montréal. Here Arno Penzias, a friend from Bell Labs who was working in radio astronomy, told him about a continuing problem with radio noise being picked up by his very sensitive Bell Lab radio telescope. When Penzias and his colleague Robert Wilson searched their data they always found a low, steady noise that persisted in their receiver. This residual noise was 100 times more intense than they had expected, was evenly spread over the sky and was present day and night.

They were certain that the radiation they detected at a wavelength 7.35 centimetres did not come from the Earth, the Sun or the Milky Way galaxy. After thoroughly checking their equipment, removing some pigeons nesting in the antenna and cleaning out the accumulated droppings, the noise remained. They concluded that this noise was coming from outside our own galaxy, although they were not aware of any radio source that would account for it. *The Bell Labs pair could simply not get rid of the unwanted signal.*

When just a few months after meeting Penzias, Burke received the draft from Peebles, he immediately recalled the conversation in Montréal about the 'noise'. He telephoned Penzias at Bell Labs straight away, for the first time making the connection between the predicted cosmic background radiation described by Peebles and the noise signal being detected by Penzias and Wilson at Bell Labs.

The penny had dropped! Penzias at last had an explanation for the mysterious noise picked up by his telescope; it wasn't noise at all, but *the radiation echo from the creation of the universe*. He immediately telephoned Dicke at Princeton to tell him that the cosmic background radiation described in Peebles' draft had been detected by the radio telescope at Bell Labs.

In a now famous and perhaps apocryphal addendum to the story, it is said that Penzias' phone call interrupted a lunchtime meeting at Princeton at which Dicke was discussing with his group the construction of the sensitive microwave receiver to detect background radiation. As the story goes, after taking the call Dicke returned to his meeting and said, 'Boys, we've been scooped!'

Now, the game was up. There was no point in continuing the Princeton project because Penzias and Wilson had already verified the prediction and confirmed the presence of Big Bang radiation. In an ironic twist, it transpired that the Bell Labs team had used a Dicke Radiometer.

The next day Dicke, Peebles and his team travelled up the New Jersey Turnpike to the Bell Labs site on Crawford Hill to inspect the radio telescope and experimental data. They were very impressed with how meticulous Penzias and Wilson had been with their measurements and how they had eliminated all possible sources of spurious signals. The Princeton people also heard how Penzias and Wilson reduced their data very carefully, yet still found a low, steady, mysterious noise that persisted in their receiver.

Here was a perfect symbiosis, one team had a theoretical prediction and no observations; the other team had reliable observations but no theoretical explanation for their measurements. In order to avoid potential conflict, they decided to publish their results in the same issue of the *Astrophysical Journal* in the summer of 1965. In the first paper, Dicke and his associates outlined how the calculation was set up and explained the importance of the cosmic background radiation as proof of the Big Bang theory. The second paper, jointly signed by Penzias and Wilson, was given the strange title of *A Measurement of Excess Antenna Temperature at 4,080 Megacycles per Second*. They omitted the cosmological implications of their discovery and simply noted the existence of an unexplained residual noise temperature. Explanation of the mysterious background noise was deferred to Dicke and Peebles in the companion paper that appeared in the same issue.

Dicke was particularly frustrated by the pre-emptive 'scoop' because it seemed certain that given a few months – even weeks – the Princeton experts would surely have detected the background radiation. Furthermore, it was thought that if Penzias had not described the troublesome noise in his antenna to Burke in Montréal in late 1964, the Bell Labs duo would have continued to be puzzled and would probably have read about the prediction and detection of the radiation in an article written by Dicke and his associates once they had made the measurement themselves. As it was, Penzias and Wilson would win the Nobel Prize in 1978 after a decade of scrutiny and refinement by astronomers from all over the world. The data revealed a precise fit to the classic thermal black-body theory. The Princeton team was not mentioned at all in the award citation.

The Sept 29ᵗʰ 1963
Gamow Dacha
785 · 8th Street
Boulder, Colorado

Dear Dr. Penzias,

~~Send~~ Thank you for sending me your paper on 3°K radiation. It is very nicely written except that "early history" is not "quite complete". The theory of, what is now knows as "primeval fireball" was first developed by me in 1946 (Phys. Rev. 70, 572, 1946 ; 74, 505, 1948 ; Nature 162, 680, 1948). The prediction of the numerical value of the present (residual) temperature ~~can~~ could be found in Alpher & Herman's paper (Phys. Rev. 75, 1093, 1949) who estimate it as 5°K, and ~~in~~ in my paper (Kong.Dansk.Ved.Sels. 27 no10, 1953) with the estimate of 7°K. Even in my popular book "Creation of Universe" (Viking 1952) you can find (on p. 42) the formula $T = 1.5 \cdot 10^{10} / t^{1/2}$ °K, and the upper limit of 50 °K. Thus, you see the world did not start with almighty Dicke. Sincerly G. Gamow.

Gamow's cynical handwritten letter to Penzias, 1963.

There is some sentiment that Ralph Alpher should have shared the Nobel award with the Bell Labs scientists as he first predicted the existence of the radiation with Gamow in 1948. Furthermore, he predicted nearly the exact characteristics of temperature and wavelength in 1949 with Herman. Gamow died in 1968 and would not have been eligible for the Nobel award in any case, which is never given posthumously. However, some measure of the degree of bitterness felt by Gamow is revealed in a cryptic handwritten note he sent to Penzias in Sept 1963 in which he reminds Penzias that his 'early history . . . is not quite complete'. Gamow then lists several references to his own work as well as to the works of Alpher and Herman before closing with the cynical comment that, '. . . you see the world did not start with almighty Dicke.'

Gamow proselytized for recognition for a few years after the Bell Labs discovery, without success, but never lost his great sense of humour. Once, when questioned by a reporter about the priority of the discovery, he responded that someone (Penzias and Wilson) 'had found the nickel he had lost in 1948'. As the main contributor to the research programme that predicted the background radiation fifteen years earlier, Alpher must have been offended by descriptions of his work that emphasized only the nucleosynthesis of the elements in the early universe and failed to consider his prediction about cosmic radiation. Dicke later wrote a popular review article on the Big Bang radiation in *Scientific American* and still failed to acknowledge Alpher.

Alpher was outspoken about his lack of recognition from Bell Labs or Princeton:

Was I hurt? Yes! How the hell did they think I'd feel? I was miffed at the time that they never even invited us down [to Bell Labs] to see the damned radio telescope. It was silly to be annoyed but I was.

The steady state team's reaction was mixed. Biondi accepted the cosmic microwave background radiation (CMB) theory and gave up on the steady state theory he had helped to create and nourish for two decades. But Gold and Hoyle were steadfast in their backing of steady state theory. Predictably, Hoyle used his genius to developing a new version of the steady state theory called the quasi-steady state (QSST) in conjunction with one of his former graduate students, Jayant V. Narlikar, who has subsequently worked as a cosmologist in India. This theory imitates the Big Bang model by postulating a universe with alternating phases of contraction and expansion, with matter being created in intense bursts. In spite of attracting a few important collaborators, the QSST theory did not survive.

While in Stockholm for the awarding of the Nobel in 1978, Penzias became philosophical and developed some generosity. Apparently, he had made peace with Alpher in a long private conversation shortly before preparing his remarks for the Nobel acceptance speech. Though he did acknowledge Alpher's earlier work on background radiation, he made the somewhat self-serving remark that none of the earlier *published* papers had made any mention of surviving relict radiation. Naturally, he wished to emphasize the work cited by the Nobel committee itself, and said:

> The theory and observation were then brought together in a pair of papers (Dicke et. al.; Penzias and Wilson 1965) which led to decisive support for evolutionary cosmology and further renewal of interest in its observational consequences.

He went on to further claim that the Bell Labs 1965 results were the proof of the validity of the Big Bang:

> The existence of the relict radiation established the validity of the expanding universe picture with its cosmological production of the light elements hydrogen, deuterium and helium during the hot early stages of the expansion. The build-up of the heavier elements occurs at a much later stage after the stars have formed. Much of the build-up of the heavier elements goes on in a few violent minutes during the life of massive stars in which their outer shells are thrown outward in supernova

explosions. This mechanism accounts both for the formation of the heavy elements as well as their introduction into interstellar space. Thus, the total picture seems close to complete and only a few puzzles remain. One thing is clear however, observational cosmology is now a respectable and flourishing science.

Thus, the Big Bang was now the only game in town. The verification of the cosmic background radiation was the last of the three critical observations that placed strict constraints on all cosmological models. Any new model from this point on would have to be as precise as the Big Bang model itself in describing the following:

- Expansion of the universe
- Primordial abundances of the light elements
- Existence and character of the microwave background radiation

Any successful cosmological theory must explain these three observations in quantitative detail, using theoretical models that agree as far as possible with known laboratory physics. Consequently, most active work in cosmology today is done within the Big Bang framework because of its continuing success in meeting these criteria, even as observational data become more and more accurate. Thus, cosmologists are forced to accept the idea of a hot, dense early universe that is relatively uniform and isotropic i.e. the same in all directions.

Challenges of the Big Bang
Since the discovery of the background radiation in 1965, and its resulting confirmation of the Big Bang theory, there have been some critics who remain sceptical. These sceptics have focused on highlighting difficulties that the theory has not explained, including the formation of galaxies and the influence of dark matter on the properties of the universe.

The question of the formation of galaxies was raised because in the preliminary measurements of Penzias and Wilson, the radiation background was so uniform that there was no evidence at all of the clumping of matter to eventually form nebulae and galaxies. This question was emphatically settled by a NASA-sponsored project, which was to become one of the most spectacular experiments in history.

Cosmic Background Explorer (COBE)

It took a decade for astronomers to devise the apparatus and experiments that were necessary to orbit around the Earth and measure the cosmic background radiation. The Bell Labs experiment had only measured the radiation at one wavelength and there were two very important questions left to answer. Was the radiation spectrum classically thermal, and observed at the temperature predicted by the theory of the Big Bang? And was there any evidence for irregularities in the radiation pattern that could then be interpreted as evidence for the formation of galaxies? These questions could prove or disprove whether a definitive measurement of cosmic background radiation had actually been made.

In 1974, NASA issued an 'Announcement of Opportunity' for astronomical missions that could use the small or medium-sized Explorer spacecraft. Out of the one hundred and twenty-one proposals received, three dealt with studying the cosmological background radiation. Though these proposals ultimately lost out to the Infrared Astronomical Satellite (IRAS), the strength of the three proposals sent a clear message to NASA: this was a matter worth investigating in the near-future. Two years later, NASA selected members from each of the three proposal teams of 1974, suggesting they get together and propose a joint project. After a year of planning, this team came up with a polar orbiting satellite that could be launched by the Space Shuttle Challenger. They called the satellite COBE, the *Cosmic Background Explorer*. However, when an explosion destroyed the Space Shuttle Challenger, NASA forbade the launching of COBE.

On 18 November 1989, the COBE apparatus was finally placed into orbit aboard a Delta rocket. COBE I used an absolute microwave radiometer calibrated by a bath of liquid helium on board the satellite. Once in orbit, the instruments took only eight minutes to verify the conclusions based on the preliminary 1964 measurements of Penzias and Wilson. This time however, measurements were taken at many different wavelengths instead of just a single wavelength. The data traced out a near perfect thermal radiation curve for a body at a temperature of 2.736 degrees above absolute zero, very close to the temperature for the background radiation that had previously been predicted by theory.

The results proved without a doubt that the detectors were sensing the remnant of the hot, dense state of the early universe. Such a curve would have thrilled nineteenth century physicists like Max Planck, who developed classical thermodynamics, as much as it did the American Astronomical Society, when the results were first presented in 1990.

But the big news was still to come. COBE II, which employed a sensitive differential microwave radiometer, didn't measure the *absolute* temperature of the radiation at a given point in the sky. Rather, it measured the *difference* in temperature between two points. If, for example, the COBE I single antenna confirmed that the temperature at point A is 2.736 degrees, the COBE II dual antenna differential radiometer recorded that the temperature difference between point A and point B is 0.002 degrees.

The goals of the two measurements were quite different, as were the two teams of scientists who planned and carried out the two distinct programmes. COBE I was the baby of John Mather of NASA. As COBE's scientific leader, Mather worked to keep a 1,500-person team focused on the science of the project. There were many hurdles, including the aforementioned tragedy of the explosion of the original launch vehicle, the Challenger Shuttle. As we have seen, this disaster sent the team back to the drawing board to redesign COBE for a later launch by a Delta rocket.

COBE II, the differential microwave radiometer that worked to determine the variations or anisotropy in the cosmic background radiation, was under the supervision of George Smoot of the University of California. Smoot's project, to look for evidence of ripples in space time – critical in answering questions about the formation of galaxies – was much more difficult to analyse. However, in April 1992, after more than two years of data collecting and examination, Smoot and his team made a dramatic announcement. The satellite had detected tiny temperature variations of the order of about 1/100,000ths of a degree in the background radiation. Smoot made the following announcement: '. . . according to computer-generated data plots of the entire sky, the temperature was minutely higher in the direction of the large galactic clusters and slightly lower in the great cosmic voids.' These ripples in the background radiation were just large enough to explain some of the structures seen in today's universe in terms of events that took place billions of years ago. The report was greeted with an enthusiastic media response all over the world.

The Cambridge cosmologist Stephen Hawking called it the 'greatest discovery of the century . . . if not of all time.'

The results of these two investigations were enough to earn John C. Mather (NASA Goddard Space Flight Center, Greenbelt, MD) and George F. Smoot (University of California, Berkeley, CA) the 2006 Nobel Prize in Physics.

This story has a poignant ending. Ralph Alpher died in August 2007, but according to the science journalist Marcus Chown, in an obituary in a London newspaper:

> . . . he died a contented man. A turning point was the launch in 1989 of Nasa's Cosmic Background Explorer (COBE) satellite. On the morning of 18 November, the scientists and engineers who had worked on the project stood about in the freezing pre-dawn darkness waiting for the launch of the Delta rocket that would carry COBE into space. Among them in the crowd, stamping their feet to keep warm, were Alpher and Herman. COBE's project scientist, John Mather, had made a special point of inviting them. At long last, everyone recognized Alpher's prescience in predicting the afterglow of the big bang.

EPILOGUE

During this journey of nearly forty centuries, to discover the universe from the Babylonians to the Big Bang and from Hipparchus to Hubble, it is not an exaggeration to claim that the very nature of science was developed. This tale inevitably reveals how the practice and perfection of scientific work has been observed, discovered, recorded and debated. This is particularly true of the period following the scientific revolution. Considering the total length of this history, during the last four centuries, beginning with the publications of Tycho, Kepler and Galileo, the rate of progress has been breathtaking. The now well-established standard model of the Big Bang has been accepted by most who believe in established science and the critical tests of cosmology have all finally been satisfied. This is an extraordinary achievement for the whole of mankind.

The art of observation has been at the heart of the quest over this vast expanse of human history. Observation has been central in the evolution of humanity's understanding of the universe. Even Einstein's universe, that based on his remarkable theory of general relativity, would never have developed without centuries of results observing the heavens to compare with the results found in Earth-bound laboratories. Clues provided by Galileo's seventeenth century experiments, and by the recorded irregular motion of the planet Mercury, became the source of Einstein's thought experiments that led to the principle of relativity, the equivalence principle and ultimately to the famous field equations.

In 2009, we look back over four hundred years to celebrate the 1609 publications of Galileo's *Sidereus Nuncius* and Kepler's *Astronomia Nova*. These two milestones, recognized by UNESCO during 2009 as the Year of Astronomy, have provided the raison d'être for books such as this one. Authors hope that readers may from time to time look up into a clear evening sky and marvel not only at the beauty of the planets, the Moon and stars, but also at the realization that a fraction of 1 per cent of the total radiation impinging on a skygazer's face at any time is coming from the residue of the Big Bang flash – still present in space after 13.7 billion years.

Though we may not be able to feel the effect of this miniscule radiation, thanks to the work astronomers on the COBE project, we are sure it is present. .Previously opaque and filled with highly energetic plasma, the universe cooled as it began to spread out, eventually becoming transparent. Inhomogeneous clumps of mysterious dark matter, imbedded in the transparent universe, then provided the seeds for the galaxies to condense. COBE has confirmed this picture down to the last detail, including the 'wrinkles' that have been interpreted as evidence for galaxy formation. In this experiment, the *human race was able to mark the extent of its own realm from an impossibly insignificant viewpoint on one planet, revolving around one star, in one spiral galaxy amongst billions of others*.

It is almost fifty years since Arno Penzias made the fateful phone call to Robert Dicke in New Jersey proclaiming the discovery of the cosmic background radiation. Since that time, many detailed measurements have been made of this radiation and, as a result, cosmology is now a respectable, experimental science. Additionally, the Big Bang model has attracted more and more scientists due to the comprehensive nature of its predictions, which agree with all observations. At this point in history, it seems unlikely that the Big Bang picture of a uniform, hot and dense, early universe will be modified very much. Though the physics of the absolute beginning of the universe is not understood – and it may never be – the nature of the growth of the universe from the period of 'inflation' (close to the start of the expansion) is consistent with the Big Bang.

Wendy Freedman of the Carnegie Institution Observatories reported at a recent international conference on the status of cosmology. Freedman, one of the major experts on all that is known about the universe, claimed that the field of cosmology is now, at the beginning of the twenty-first century, free of serious controversies. Can this be true?

Yes, at this stage there is general quantitative agreement by different teams of observers around the world on the value of the elusive Hubble constant, which supports a consistent value of the age and the size of the universe. Furthermore, there is also a consensus that the gravitational attraction of cold dark matter provides a successful explanation of how large-scale structures like galaxies were formed. With astounding confidence, Freedman and her colleagues state that ordinary matter makes up only 4 per cent of the total mass energy of the universe and that dark matter and dark energy make up the other 96 per cent.

Not surprisingly, her report also confirms the validity of the Big Bang model in which the universe evolved from a hot, dense primordial phase expanding in a manner described by models based on solutions to Einstein's equations of general relativity. It is fitting that scientists from the Carnegie Institution announced these results. It was G.E. Hale, supported generously by Andrew Carnegie, who first took astronomy high into the California Mountains in the early part of the twentieth century; and it was here that many historic discoveries were made about the age, size and growth of the universe.

Discoveries will continue to be made in the twenty-first century, but those who are interested in the overall structure of the universe should feel privileged to live during what has been truly the golden age of astronomy – or rather, cosmology. This will only happen once!

FURTHER READING

PART I: Ptolemy's Universe

Crowe, Michael J. *Theories of the World from Antiquity to the Copernican Revolution*. New York: 1990.

Evans, James *The History and Practice of Ancient Astronomy*. Oxford University Press, USA: 1998.

McEvoy, J. P. *Eclipse: Science and History.* Fourth Estate: 2000.

Neugebauer, O. *A History of Ancient Mathematical Astronomy*. Berlin: 1975.

Ptolemy, *The Almagest*, Trans. G.J. Toomer. New York: 1984.

Rubenstein, Richard E. *Aristotle's Children: How Christians, Muslims, and Jews Rediscovered Ancient Wisdom and Illuminated the Middle Ages*. Harvest Books: 2004.

PART II: Newton's Universe

Caspar, Max *Kepler*, Trans. C. Doris Hellman with notes by Owen Gingerich and Alain Segonds. New York: 1993

Cohen, I.B. 'Newton's Discovery of Gravity'. *Scientific American*, March 1981.

Copernicus, Nicholaus *De revolutionibus orbium coelestium (On the revolutions of the heavenly spheres)*, Trans. Edward Rosen. Baltimore: 1978.

Donahue William H. (ed.), *Selections from Kepler's Astronomia Nova*. Green Lion Press: 2004.

Drake, Stillman *Galileo at Work: His Scientific Biography*. Chicago: 1978

Galilei, Galileo *The Starry Messenger*, Trans. A. van Helden. Chicago: 1989.

Gingerich, Owen *The Book Nobody Read: Chasing the Revolutions of Nicolaus Copernicus.* Arrow Books Ltd: 2005.

Gingerich, Owen *The Great Copernicus Chase and other adventures in astronomical history.* Cambridge: 1992.

Newton, Isaac Dana Densmore, William H. Donahue (Trans.), *Newton's Principia: The Central Argument. Translation, Notes, and Expanded Proofs.* Green Lion Press, 1996.

Reston Jr, James *Galileo: A life.* Harper Collins, New York: 1994.

Stephenson, Bruce *Kepler's Physical Astronomy*. Princeton University Press: 1994.

Thoren, V. *The Lord of Uraniborg: A Biography of Tycho Brahe.* Cambridge: 1990.

PART III: Einstein's Universe

Alpher Ralph A. & Robert Herman, *Genesis of the Big Bang.* Oxford University Press, USA: 2001.

Bais, Sander *Very Special Relativity: An Illustrated Guide*. Harvard University Press: 2007.

Chown, Marcus *Afterglow of Creation: Decoding the Message from the Beginning of Time*. University Science Books, USA: 1996.

Christenson, Gale E. *Edwin Hubble: Mariner of the Nebulae.* University of Chicago Press: 1996.

Gamow, George *The Birth & Death of the Sun: Stellar Evolution and Subatomic Energy.* Dover Publications: 2005.

Gamow, George *The Creation of the Universe*. Dover Publications: 2004.

Hirshfeld, Alan W. *Parallax: The Race to Measure the Cosmos.* Holt: 2002.

Hubble, Edwin *The Realm of the Nebulae*. Yale University Press: 1982.

Johnson, George *Miss Leavitt's Stars: The Untold Story of the Woman Who Discovered How to Measure the Universe.* W.W. Norton & Co.: 2005.

Kragh, Helge H. *Cosmology and Controversy: The Historical Development of Two Theories of the Universe.* Princeton University Press: 1996.

Longair, Malcolm S. *The Cosmic Century: A History of Astrophysics and Cosmology.* Cambridge University Press: 2006.

McCray, Patrick W. *Giant Telescopes: Astronomical Ambition and the Promise of Technology.* Harvard University Press: 2006.

Mitton, Simon *Fred Hoyle: A Life in Science.* Aurum Press Ltd: 2005.

Pais, Abraham *Subtle is the Lord: The Science and the Life of Albert Einstein.* Oxford University Press: 1982.

Peebles, James P. et. al., 'The Evolution of the Universe'. *Scientific American*, October 1994.

Singh, Simon *Big Bang, The Most Important Discovery of All Time and Why You Need to Know About it.* Harper Perennial: 2005.

Stanley, M. *Practical Mystic: Religion, Science, and A. S. Eddington.* Chicago University Press: 2007.

General

Ferguson, Kitty *Measuring the Universe.* Walker: 2000.

Ferris, Timothy *Coming of Age in the Milky Way.* Harper Perennial: 2003.

Freedman Wendy (ed.), *Measuring and Modeling the Universe*, Vol. 2. Cambridge University Press: 2004.

Hoskin Michael (ed.), *The Cambridge Illustrated History of Astronomy.* Cambridge: 1997.

INDEX